SCIENTIFIC AMERICANS

INTERSECTIONS IN LITERATURE AND SCIENCE

Since the turn of the century the study of literature and science has been among the fastest-growing and most innovative areas of literary and historical research. Through the application of rigorous interdisciplinary scholarship, studies in literature and science have offered a keen appraisal of the relationships between the historical emergence and significance of the sciences, as well as the literary and artistic cultures that engage with and critique them. The field has recognized the importance of sustained and detailed historical research whilst maintaining a close regard for the literary imagination. At the same time, critics and scholars have understood that broader philosophical questions arising from the study of literature and science must also be addressed and have actively sought to develop the philosophical implications of the intersections and tensions between the two disciplines without negating the importance of their social, cultural and political contexts. This series aims to promote the research and scholarship of literature and science's keenest advocates and most talented critics. The studies in the series, unrestricted by period, locale, or genre ,will offer fresh insight into the intellectual history of literary texts and scientific developments, and in doing so will advance the central paradigms of literature and science scholarship whilst enhancing and developing the field's methodological practices.

SCIENTIFIC AMERICANS

THE MAKING OF POPULAR SCIENCE AND EVOLUTION IN EARLY-TWENTIETH-CENTURY U.S. LITERATURE AND CULTURE

John Bruni

UNIVERSITY OF WALES PRESS
CARDIFF

www.uwp.co.uk

British Library CIP Data
A catalogue record for this book is available from the British Library.

ISBN: 978-1-7831-6017-4
e-ISBN: 978-1-7831-6018-1

Printed by CPI Antony Rowe, Chippenham, Wiltshire

CONTENTS

ACKNOWLEDGEMENTS

Writing this book has been far from an individualistic effort. I would like to thank the following people for their guidance, enthusiasm and support: Cheryl Lester, Philip Barnard, Elizabeth Schultz, Susan Harris, David Katzman, Janet Sharistanian, Mary Greenberg, Bruce Clarke, Dale M. Bauer, Karla Armbruster, Carol Colatrella, Joseph Tabbi, Stacy Alaimo, Judy Sneller, Alfred Boysen, Lisa Yaszek, Philip B. Sharp, Doug Davis, Michael Bryson, Karl Zuelke, Janine DeBaisse and Christopher Phelps.

Thanks especially to Ursula K. Heise and Barry Shank for allowing me to quote from unpublished work.

The librarians and archivists at Massachusetts Historical Society, Huntington Library (San Marino, CA), University of Kansas and Grand Valley State University offered invaluable help in finding sources.

Earlier versions of some of the book chapters were published elsewhere. My reading of Jack London's *The Call of the Wild* in chapter 3 first appeared in 'Furry logic: biological kinship and empire in Jack London's *The Call of the Wild*', *ISLE: Interdisciplinary Studies in Literature and Environment*, 14/1 (Winter 2007), 25–49. An earlier version of chapter 3 was published in 'Performing the perfect dog: the reconstruction of gender in Jack London's *The Call of the Wild* and *White Fang*', in Bernadette H. Hyner and Precious McKenzie Stearns (eds), *Forces of Nature: Natural(izing) Gender and Gender(ing) Nature in the Discourses of Western Culture* (Newcastle upon Tyne: Cambridge Scholars Press, 2009), pp. 174–210. An earlier version of chapter 4 was published in 'Becoming American: evolution and performance in Edith Wharton's *The Custom of the Country*', *Intertexts*, 9/1 (Spring 2005), 43–61. An earlier version of chapter 5 was published as 'The miseducation of Henry Adams: fantasies of race, citizenship, and Darwinian dynamos', in Jeannette Eileen Jones and Patrick B. Sharp (eds), *Darwin in Atlantic Cultures: Evolutionary Visions of Race, Gender, and Sexuality* (New York and London: Routledge, 2010), pp. 260–82.

I am indebted to those at the University of Wales Press who helped the project come to fruition: Alan Rauch, Martin Willis, Sarah Lewis and the (unnamed) manuscript reader.

Finally much gratitude to my family. And to its non-human animal members, Marley and Clyde.

And, last – and most crucially – Rachel Anderson, who has contributed more than words could ever express.

This work is dedicated to the memory of my mother, Barbara Bruni.

LIST OF ILLUSTRATIONS

INTRODUCTION

Founded in 1845, *Scientific American* was the first magazine to disseminate scientific knowledge about the world in concise, journalistic prose, gleaned from the perspectives of specialists in a variety of academic disciplines, to a 'general' audience. *Popular Science Monthly* followed in 1872, seeking the readership of 'educated laymen' and offering in-depth articles from well-known authorities in their fields such as Herbert Spencer (in philosophy), T. H. Huxley (in biology), William James (in psychology) and John Dewey (in education). Both magazines let the reader listen into a multi-dimensional, far-ranging and provocative conversation not limited by disciplinary constraints.[1] Every article sheds light on the times in which it was written, carrying a piece of a larger worldview that sets the terms for its writing. Article by article, the tone varies: some are more confident about reaching a consensus on an issue, others more pessimistic. In a sense, this is science in the making, in action.

But in no way is science in the making beyond the marketplace of ideas. While the combination of tutorials in elementary science and reports on the technical details of the latest military campaigns proved a continuing windfall for *Scientific American*, by 1900, *Popular Science Monthly* was in financial distress, its forays into the abstractions of theoretical scientific thinking unable to attract a sufficient number of readers. By 1915, it had been sold and reformatted to look much more like its rival.[2] Not only do the different outcomes of *Scientific American* and *Popular Science Monthly* help to illustrate the widespread influence of economics, albeit perhaps on a small scale; a focus on how a growing national economy animated desires for global imperial adventures overseas was also shared by both magazines. In the early twentieth century, *Scientific American* frequently ran reports that linked national dominance in global trade to the acquisition of foreign territories and those likely to be acquired, while *Popular Science Monthly* mapped the international dimensions of the economic benefits of scientific research. Thus from the viewpoint of these magazines, scientific practice became deeply enmeshed in an early-twentieth-century shifting worldview, informed by debates about immigration, national expansion and technology.

From 1895 to 1910, one of the dynamic components of this worldview was evolution. Most accurately depicted, evolution comprises multiple theories about the influence of heredity and environment and natural selection: individuals with traits that help them successfully adapt to their environment pass these traits on to their offspring. These theories were codified, explained and disseminated by Charles Darwin. Questions arose from a variety of disciplines about the validity of Darwin's model of evolutionary development. Popular science journalism in particular challenged the idea that there was widespread agreement about how evolution worked. In doing so, it points out that the cultural implications of evolution were very much in play, for while it was understood that humans were part of, not separate from, evolutionary processes that shaped the natural world, there were clearly limits on the influence of what is known as social Darwinism, most often defined as a doctrine stating that cultural competition mirrors the struggle for survival in the natural world. A survey of *Scientific American* and *Popular Science Monthly* provides crucial evidence for how the doctrine in the US failed to have a complete monopoly on the cultural meanings of evolution, contra to Richard Hofstadter's famous argument that social Darwinism permeated American thinking, and sets up a critique of social Darwinism.[3] This critique states that social Darwinism overly credited either Darwin or Herbert Spencer (or sometimes both) for the cultural readings of evolutionary theories, an assumption that overlooks the fluid interplay of scientific and cultural tropes that creates a thick intellectual context for evaluating evolution. Furthermore, the social Darwinist narrative relies on preformed concepts such as survival of the fittest and natural selection, without addressing how these concepts are formulated over time through interdisciplinary discussions, for instance about the cultural construction of race, class and gender categories, that reappraise the connections between scientific and social realms. Popular science journalism pushed against the monolithic ideology of social Darwinism, suggesting that evolutionary theories could be situated on a far-ranging political spectrum that goes beyond social Darwinist doctrine. In sum, it pointed out the dangers of the tendency to make simplistic references to this doctrine, for such a tendency fails to account for differing political ideologies and the variety of thinking about evolution. To go beyond social Darwinist models for understanding the cultural meanings of evolutionary theories in the early twentieth century, we can, following popular science magazines, consider the range and scope of the scientific debates about biological processes that speculate upon whether survival of the fittest and natural selection are random and to what degree they guide evolution.

Such debates, moreover, energized cultural dialogues about the consequences of immigration and expansion, both of which continuously raised the question: who is/can/will be an American? In the wake of legal ambiguity over a series of Supreme Court cases and public reaction (anti-immigration and anti-imperialism movements), there was increasing pressure for scientific definitions of citizenship.[4] The question of how to qualify the role of evolutionary processes in shaping domestic and foreign affairs, however, remained unresolved. What clues, if any, did evolution hold for the national future – a promise of unlimited progress or the threat of inevitable decline?

This question, in particular, fascinated several of the most scientifically literate US writers, namely Theodore Dreiser, Edith Wharton, Jack London and Henry Adams, who all left a considerable mark on the early-twentieth-century national imagination.[5] Afforded by the expanded scientific context enabled by popular science journalism, a close examination of their writing illustrates that, rather than getting their ideas about evolution second-hand, filtered through a social Darwinist ideology, they actively determine what evolution means. Their working definitions of evolutionary theories emerge from their dramatic exploration of the possibilities and limits of biological models for individual and collective self-making.

In turn, their focus on biological models reflects the scientific basis for naturalism, a philosophy they all loosely share.[6] Viewed through the perspective of popular science journalism, their naturalist affiliations, however, become tested. None of the writers, of course, would have disagreed with naturalism's insistence that scientific principles guide human behaviour and development. Nor would they have questioned the belief that humans are subject to biological and cultural influences beyond their immediate control. Indeed, they helped to clarify and refine naturalism's emphasis on the evolutionary principles of natural selection, heredity and environment, as well as sharing naturalism's interest in anthropology and the social sciences as having the potential to expand evolutionary thinking. Yet all of the writers' treatments of evolutionary processes, at one time or another, question an oversimplification of evolutionary theories, which naturalism often tended to make by emphasizing only the best-known and most readily comprehensible features of evolution.

These writers' attempts to capture the complexity and contingency of evolution line up with the science in the making depicted by popular science journalism. Writers and journalists both illustrated how academic disciplines opened up evolution to new possibilities (such as thermodynamics). And like popular science magazines, these writers interrogated social Darwinist doctrine (such as survival of the fittest), but they went much further by giving evolutionary thinking a dramatic trajectory that sheds light on some of the more profound implications of evolution. For one, the writers' reformulation of the naturalist plot of decline illustrates how the failure to survive rested on a troubling logic: individual effort to avoid extinction is overshadowed by a larger background of biological inheritance. Hence self-initiative, as a competitive ethos underlying social Darwinism and as a larger foundation of self-awareness, is undercut.[7] A reading of their narratives also shows how hierarchies of race, class and gender guided attempts to depict scenarios of natural and social progress. By imagining how bodies – both human and non-human – undergo evolutionary transformations, these writers pointed out that evolution destabilized identity. Furthermore, the way in which these writers played with the naturalist narrative reveals the profound ambivalence that undergirded evolutionary fantasies about natural/national development.

These writers were prepared to assert that there were no easy answers to evolutionary questions about identity (at the same time drawing attention to the instability of identity as constructed through race, class and gender categories). How they construct the possibilities and limits of global citizenship questions the idea of universal definitions for race, class and gender. Their discussions about evolution explore the performance of identity through collective and entangled bodies always under construction, always *becoming, not simply being*.

In fashioning new ways of looking at evolutionary becomings, they prompt us to align their narratives with a vision of emerging global ecosystems in the early twentieth century. Surprisingly, there has been little effort to examine these writers from an ecocritical perspective, a perspective that would change how we view both citizenship and our planetary obligations. For example, questioning the myth of the autonomous self puts limits on the individualistic will of political leaders that informs policies of global acquisition and aggression. Recognizing that evolution has produced us as only one part of global life relations undermines the belief in human superiority. Furthermore, the ways that organisms are connected to and yet remain apart from their environments parallel the conditions under which national subjects are and are not global citizens. For instance, while the identity of the American abroad is shaped by the idea of a global selfhood, that idea is animated by US imperialist desires that, projected onto the world stage, privilege national interests and restrict citizenship.

While evolution shapes and is shaped by a shifting world picture, such evolutionary transformations take place against an even more fundamental change in observation – a change that has until now received at best marginal commentary. Granted, both popular science journalism and the writers discussed here, as we have noticed, depict scientific practice as embedded in a cultural milieu. That depiction discloses significant limits on scientific observation as completely objective truth-telling, which leads us to a rather remarkable observation: science and the arts are speaking not different but *similar* languages. Just as arriving at any unified perspective of evolution was a complicated and messy process, *Popular Science Monthly* in the early twentieth century published a number of articles on art, claiming that the internal complexity of an artistic work is built up through its being observed, from the reactions of its observers. The stage is set for an important shift: from first- to second-order observation. What occurs is a pressuring of simplistic notions of first-order observation, that is, an observer observing objects. What emerges is a second-order model of observation, where observers watch other observers. Dreiser and Wharton in particular play with the idea of second-order observation in art, where the focus is on how an artwork fashions a social space for observers to watch other observers.

Over time, theories of artistic observation emerge from the scientific consideration of complex systems. While the subsequent book chapters will discuss systems theory in more detail, its foundational aspects may be described thusly. To describe how complex systems work, cognitive biologists Humberto Maturana and Francisco Varela create the concept of autopoiesis (self-making), which folds observation into a schema of self-reference.[8] This holds that the process of self-reference through which systems maintain themselves makes these systems both connected to and apart from their environments; they therefore cannot be said to perceive their environments directly. Autopoiesis is closely linked to second-order observation because of self-reference: systems cannot get objective knowledge from their environments, for such knowledge can only be obtained from another system that is blind to its own limited vision observing the first, and so on. In the literary texts I examine, the uncertainty of vision that denies a stable point of reference constitutes the unbounded and unstable identities in Dreiser, Wharton and London. Extended to Adams, this principle impels him to question the validity of his own historical reporting.

We may then turn to systems theorist Niklas Luhmann, who writes about both psychic and social systems – the latter including art.[9] His elaboration on Maturana and Varela's argument discloses three important points that draw together the arts and sciences through a system/environment distinction. First, systems are connected to their environments through the blind spot of their vision. Luhmann's comment, which will be further explored in this study, 'reality is what one sees when one does not see it', means no one can claim to know a universal truth about the environment per se. Our perception of environmental issues depends upon the contingency of our second-order observations. Second, systems respond to environmental complexity by becoming functionally differentiated, that is, by creating subsystems that reiterate the system/environment distinction: systems become environments for their subsystems. As a result, no act of observation can take place outside a system, nor can any act of observation view the operations of all systems at once. Third, psychic and social systems co-evolve (or, as Luhmann puts it, systems interpenetrate one another), an idea that connects the biological and cultural.

I proceed to show how these three points run through naturalist writers' unfolding of evolutionary systems of biological life. For Dreiser, Wharton, London and Adams, complexity and contingency are central to the depiction of overlapping biological and social environments. Both a natural and national economy emerge through the intermingling of functionally differentiated systems, where insufficient complexity leads to system breakdown. The idea of identity, as an unstable sedimentation of race, class and gender, is rebuilt through a system's interaction with its environment. All of the writers discussed in the book, moreover, explore the contingency of second-order observation, questioning the supposed detachment of all observers – whether scientists, historians or reporters – and their claims for objective truth-making. My discussion points out that the contingency of vision disrupts yet at the same time insists on the representation of difference, which has considerable effects on the treatment of race, gender, and citizenship in the early twentieth century. In the literary narratives of Dreiser, Wharton, London and Adams, race, class and gender become exclusionary categories used to pressure the issue of who can be seen or imagined as a US citizen.

The opening chapter, 'Popular Science, Evolution, and Global Information Management', critically examines the cultural pressures, discussed here, that destabilize a pre-formed consensus about the social meanings of evolutionary theories. Such pressures are prominently represented in the magazines *Scientific American* and *Popular Science Monthly*. My discussion of these magazines, from 1895–1910, shows their chronicling of the development of new academic disciplines at the turn of the twentieth century, such as anthropology, that question how and to what degree evolution can be seen as revealing truths about human development. At the same time, national expansion and immigration compel scientific reports, addressed to a popular audience, about the biological origins of citizenship and overseas regions as possible sites for economic and political control by the US. These reports, I elaborate, are shaped by evolutionary ideas about progress and popular understandings of race, class, gender and national belonging. The chapter addresses how the editorial policies and practices of *Scientific American* and *Popular Science Monthly* created a concept of popular science which enables naturalist narratives about evolution to be read in a larger scientific context. Such a reading is undertaken in the following chapters.

Chapter 2, 'Dirty Naturalism and the Regime of Thermodynamic Self-Organization', investigates the question of whether the social Darwinist trope of survival of the fittest (and failure to survive as extinction) can be articulated through images of biological struggle. I trace the response to this question, which circulates through the interconnected realms of the scientific and social, through the rewritings of the evolutionary narrative of decline in Dreiser's *Sister Carrie* (1900) and Wharton's *The House of Mirth* (1905), both popular novels. My reading of their works suggests that survival depends, to a large degree, on the connection of economics to metabolic processes. In other words, it is a model that ties economic flows to energy flows. I consider how the thermodynamic language that informs this model also shapes the emergent 'new economy' in the early twentieth century that promotes stock market speculation based on the circulation of information. My argument focuses on how Dreiser and Wharton's representations of evolution suggest that the wasteful practices of Wall Street speculation are echoed by the effects of over-consumption on the natural environment.

The third chapter, 'The Ecology of Empire', situates environmental issues within the process of human and animal identity formation. London, whose natural studies in his popular novels, *The Call of the Wild* (1903) and *White Fang* (1906), counterpoint the urban perspectives of Wharton and Dreiser, makes a significant endorsement (that strongly reacts against the dominant cultural view of human superiority) of biological kinship between humans and dogs, showing their co-evolution. Yet he cannot get beyond a humanist ideology that rationalizes turning the natural landscape into a resource for economic expansion. I argue that although London depicts dogs as evolutionarily unstable and their identities open to change, London's attempts to establish an ethics for biological kinship are limited by a reliance on the legal discourses that underwrite US global imperialism. At the same time, the depiction of overseas regions as the next frontier for masculine fantasies of conquest is haunted by the question of Native American citizenship, a question that threatens to disrupt the reconstruction of a national identity.

The fourth chapter, 'After the Flood: Performance and Nation', discusses a challenge to the belief that evolution supports a biologically determined definition of identity. Returning to the work of Wharton, I examine this idea in her later novel, *The Custom of the Country* (1913). The novel takes up this challenge through her careful – yet rather radical – reading of evolutionary theory, which shapes a model of identity as not simply being but an ongoing and unpredictable process of becoming. That identity remains incomplete suggests that there can be no stable point from which observations can be made. I explain how the performance of gender roles in the novel undercuts the validity of objective reporting and thus disrupts the logic that underwrites the exploitation of both labour and nature. As I point out, Wharton suggests that the freedom of movement of US citizens is secured through unequal labour practices that shape global trade. In addition, the shift from self-possession to possession of natural resources, Wharton notes, guides how national citizenship becomes global. As the US becomes a global power, Wharton shares London's concerns about the ambivalence of citizenship for Native Americans, while relating the question of national identity to the ambiguity of race in the US constitution.

The fifth chapter, 'The Miseducation of Henry Adams: Fantasies of Race, Citizenship and Biological Dynamos', explores how evolution can be re-envisioned as a process of increasing complexity that redefines ideas of chaos and order, unity and multiplicity. Adams, whose idiosyncratic worldview distinguishes him from the writers discussed in the previous chapters, is fascinated by these possibilities for evolutionary transformations. By examining *The Education of Henry Adams* (1907), 'The Rule of Phase Applied to History' (1909) and 'A Letter to American Teachers of History' (1910), I analyse how the effects of immigration and national expansion foreground the role of self-reflexivity in Adams's historical narratives. Continuing my discussion of thermodynamics from the second chapter, I show that Adams's scientific theory of history, by connecting energy and economic flows, rationalizes the imperialistic foreign policies of the former US Secretary of State and Adams's friend, John Hay. I then examine how Adams's narration of Hay's efforts to position the US as a global leader makes explicit Adams's criticisms of Darwin. For Adams, natural selection cannot adequately explain the formation of a new definition of citizenship, what he calls the 'new American'. Furthermore, Adams's fantasy of national belonging, I demonstrate, relies on race as both a biological limit for citizenship and an abstract force that animates Adams's idea of women as 'biological dynamos'.

The concluding chapter reflects on the central goal of the project: to resituate the prose narratives of Dreiser, Wharton, London and Adams within a crucial moment when evolutionary thought and a shifting picture of the world destabilized social categories of race, class, gender and citizenship. I then propose that these writers speak beyond such a moment, for their thinking is unmistakably relevant to current popular and scientific debates about global and environmental imperialisms – debates that rethink the future of technoscientific empire and the remaking of selfhood.

1

POPULAR SCIENCE, EVOLUTION AND GLOBAL INFORMATION MANAGEMENT

Published in *Scientific American* in 17 October 1896, 'The development of Africa' recounts Henry M. Stanley MP's thoughts about England taking the lead in establishing a new imperialistic phase: 'Her efforts for some years after the Berlin Conference had been confined to reserving spheres of influence, rather than to violent annexation, and to moderating the passion for African land manifested by Germany, France, and Italy.'[1] The concept of 'spheres of influence' abstracts foreign policy; it encapsulates how nations take control of foreign markets through a sophisticated deployment of rhetorical posturing, diplomatic manoeuvring and statistical reporting. As I shall argue, popular scientific journalism contributes to the performance of this new phase of imperialism through its role as an observing system. The making of popular science demonstrates how information begins to have its own intrinsic value – as a political commodity – that augments the imperialistic (re)mapping of peoples and regions. At the same time, the feedback loops that connect race, gender and class formations to emergent forms of global citizenship also tie scientific reporting to strategies for global information management. We will start by questioning the separation between the social and scientific, a critical move for historicizing science that provides a new perspective on scientific rationales for national imperialist global adventures.

I. Reconstructing the social and scientific

The construction of the idea of popular science calls attention to how a scientific truth claim can be interpreted as a performative statement, or as Ira Livingston remarks, 'one that displaces both reference and self-reference. What kinds of language games is the statement playing? What does it *do*, and with and against what other kinds of statements does it operate?'[2] We might say that the power to render observation as a permanent, rather than contingent or partial, declaration of authority makes that observation socially meaningful or valuable. To put it slightly differently, as Karen Barad does,

> Performativity, properly construed, is not an invitation to turn everything (including material bodies) into words; on the contrary, performativity is precisely a contestation of the excessive power granted to language to determine what is real. Hence, in ironic contrast to the misconception that would equate performativity with a form of linguistic monism that takes language to be the stuff of reality, performativity is properly understood as a contestation of the unexamined habits of mind that grant language and other forms of representation more power in determining our ontologies than they deserve.[3]

From this perspective, I want to examine popular science journalism performatively: looking at what it does through its self-professed strategy of selection. For example, the stated objective of *Popular Science Monthly* is to be 'popular in the sense that it is not special or technical, not in the sense that it makes an appeal to all the people' and 'intended for those having a cultivated and intelligent interest in the advancement of science'.[4] The first part of the statement sets the journal's level of selectivity; it aims for neither an exclusive nor a universal audience. The second part extends the principle of selection through a performative act: possessing 'a cultivated and intelligent interest in the advancement of science' creates a class distinction. In other words, you stand apart from the general populace if you display a 'genuine' interest (that has no evaluative criteria as such) in science. Yet such a distinction does not come before or after the objective; popular science and class are co-constituted in the act of self-declaring an interest in science.

Both *Scientific American* and *Popular Science Monthly*, through their editorial choices, define popular science in part by what the term restricts or excludes. For example, from 1896 to 1899, *Scientific American* featured articles on science and sensationalism, science and spiritualism and the fraud of perpetual motion, thus attempting to distinguish popular science by demonstrating what it is not: sensationalism, spiritualism or fraud. Again the principle of selection comes into play. At this point, though, the principle is contextualized within first-level observation: X is either popular science or not, and the observer as science journalist, having un-mediated access to reality, makes the decision. The larger point, if we focus on system theorist Niklas Luhmann's comments about second-order observation (the observation of observation), is that 'the mass media create the illusion that we are first-order observers whereas in fact this is already second-order observing'.[5]

The debate about N-rays, a form of radiation supposedly discovered by Rene Blondlot in 1903, pressures this 'illusion'.[6] In the 4 June 1904 edition of *Scientific American*, an unsigned editorial, 'The N-rays: are they real or illusory?', muses: 'French eyes are certainly blessed with no greater clarity of vision than those of Englishmen; and yet the fact remains that Blondlot's rays, or at least the most important phenomena of which they are the cause, have never been observed by any but . . . French investigators.'[7] Nationality does not come before or after observation (there is no *biological* difference, strictly speaking, between French and non-French eyes), but is constituted *through* observation: what you see – or do not see – creates national differences. Nowhere perhaps is the contingency of observation made clearer; as Luhmann counsels us: 'Everything becomes contingent whenever *what* is being observed depends on *who* is being observed.'[8] But the proposed resolution of this controversy, in a rather fantastic moment, gives scientific knowledge production the ability to exceed disciplinary limits. Here, it seems, chemistry becomes conflated with physics:

> [B]y means of the old periodic law of chemistry it was possible to tabulate the chemical elements according to their properties and their atomic weights in a sequence that brought out their relation to one another strikingly. Wherever gaps occurred, it was reasonable to infer that they would be filled by elements still to be discovered – an inference that was more than once justified. By a similar tabular arrangement the N-rays may be shown to fill a gap in the series of undulatory rays.[9]

As Cary Wolfe explains, the desire to transcend disciplinary 'closure' is part and parcel of the humanist framework, whose default setting is objectivity. So this resolution would be challenged by transdisciplinary thinking, that is, a strategy that maintains the differences among disciplines. It is 'a kind of distributed reflexivity' informed 'by the fact that (by definition) *no* discourse, no discipline, can make transparent the conditions of its own observations'.[10] Transdisciplinarity reflects Luhmann's idea that observations are contingent upon those doing the observing; observations are embedded in the dynamic relationship between observer and environment. And to resist making observation transparent, to take contingency seriously, as it were, marks the critical observing process, central to any and all academic disciplines, as being of second- rather than first-order status.

By upholding its promise to serve those who have a 'cultivated and intelligent interest in the advancement of science', *Popular Science Monthly* also questions the fantasy of media as first-order observation and troubles the notion of objectivity through its commentary on scientific knowledge production. For example, we may refer to John M. Coulter's 'Public interest in research'. Published in August 1905, the article claims that the dissemination of research to the public currently happens through newspapers and magazines, and is 'scant in amount, sensational in form, and usually wide of the mark'.[11] Coulter looks at the possibilities of investigators writing their own articles for newspapers and magazines. He suggests that if the significance of the research were highlighted, the public would take a greater interest. As a result, the research would be deemed as 'practical' and then 'secure endowment for research'.[12] 'We have behind us', Coulter insists, 'a public more prosperous and much more generous, accustomed to support liberally what it is interested in. If this can be taken advantage of, there is no reason why research in America cannot be developed to an extent that is without precedent.'[13]

In March 1908, *Popular Science Monthly* reprinted Gordon Webster's address at Clark University on 1 February 1907. The address, 'America's intellectual product', claims,

> It is frequently supposed that the American public is extremely interested in the results of scientific progress, and so it is, in a certain sense. Certainly we cannot accuse it of lack of alertness, when it reads more than any other – in the newspapers. It reads with eager interest, and with implicit credulity, accounts of the supposed discoveries of science, taking at equal value the productions of notorious charlatans and those of real investigators.[14]

Part of the problem, Webster thinks, is that newspapers pay little attention to science. Few newspapers have on their staff writers who regularly 'acquaint the public with the current achievements of science'.[15] He recounts his difficulty finding the names of Nobel Prize winners, surmising that the public at large has the same difficulty. Indeed, until the US President was a recipient, he believes it likely that few even knew that such awards existed.

Webster alludes to larger modes of cultural production, art and literature, in which the nation is also behind in the competition for international prominence. He pits such productions, mentioning the names of respected authors William Dean Howells and Henry James, against the popularity at home of 'new religions, intellectual fads, and isms, ologies, and pathies of every sort'.[16] But despite such distraction, Webster says, the public is becoming aware that scientific research has 'practical applications'; it is necessary for industrial development and 'commercial dominance' of global markets.[17] Knowledge making becomes part of national identity: that 'deep thinking' is 'characteristic of Germany' leads Webster to propose that the US should make a name for itself in a similar way: 'The national government provides richly for the education of those who devote their lives to her defence; is there any less reason for providing for those who are to make her intellectually great?'[18]

Barad, in my opinion, expresses the idea that the scientific and social are coterminous most precisely:

> The point is not that there are leaks in the system where social values seep in despite scientists' best efforts to maintain a vacuum-tight seal between the separate domains of nature and culture. Nor should we conclude that the quality of the results is diminished in proportion to the permeability of this barrier. This kind of thinking mistakenly reifies culture and nature and gender and science into separate categories. But the fact is that the world isn't naturally broken up into social and scientific realms that get made separately. There isn't one set of material practices that makes science, and another disjunct set that makes social relations; one kind of matter on the inside, and another on the outside. The social and the scientific are co-constituted. They are made together – but neither is just made up. Rather, they are ongoing, open-ended, entangled material practices.[19]

Barad proposes that there is no longer a line between an observer and what is being observed; she concludes there is no longer really an outside world at all. To express this idea in the language of systems theory, as Wolfe does, '[I]t is not an "outside" to the system that is given as such but it is rather the outside of a specific inside'.[20] Thinking with Barad and Wolfe, there is not a universal outside/inside binary that would privilege the inside, and in so doing limit our agency by making subject formation dependent on such a distinction.[21]

II. Scientific and cultural narratives of expansion

The question of how the concept of popular science is created through the editorial policies and practices of *Scientific American* and *Popular Science Monthly* enables us to make a critical challenge to how narratives about evolution are often traditionally regarded. That is, to read these narratives in a larger scientific context challenges their apparent status as transparent, first-order observations. And that is important because to concede that such narratives are of first-order status becomes a bad faith agreement that, as will now be discussed, supports national imperialist global adventures.

In the traditional social Darwinist model, natural selection justifies capitalist competition. Yet such a model reduces the diversity of thinking about evolution, as natural selection comes to be regarded by those on the conservative end of the scientific political spectrum as the exclusive biological motor. An overemphasis on natural selection misrepresents why evolutionary theories build up cultural visibility. Not only are these theories widely disseminated because they appear to provide a scientific rationale for social practices; they also benefit from acting like the practices they reinforce. That is, rather than depending upon the victory of natural selection over competing principles, evolutionary theories offer a rather loose, broad-based view of biological life, a view easily compatible with a multitude of academic disciplines and cultural beliefs. These theories assimilate with other ways of thinking, much as dominant political and economic ideologies do. Livingston explains, with regard to the constant debate about the role of natural selection in guiding life,

> [F]ar from constituting a failure of evolutionary thinking, this [debate] seems to be part of the ongoing success of principles of evolution and natural selection, which continue to be incorporated in the paradigm pantheon of more and more fields, such as physics and cosmology. In other words, the depth of the paradigm (its position as exclusive center or anchor point) is being displaced as it achieves greater *reach and*

> *saturation* of other realms; its pluralization and relativization make it more resilient even as they may in some ways make it less *itself.* Not coincidentally, this seems to have been one of the leading principles of empire building for several millennia, just as it makes a rough-and-ready account of the current transnational hegemonic success of capitalism and the West.[22]

Livingston is worth quoting at length because of how radically he departs from the received wisdom about the relationship between science and culture. Evolution, in his view, does not wield a considerable cultural influence due to its victory in the economic marketplace of ideas; nor does it follow Raymond Williams's template for cultural cycles (emergent, dominant, residual).[23] Rather evolutionary doctrines gain popularity through allowing themselves to be co-opted: like capitalism, they offer no resistance to being transformed. To put it in terms of Williams's thinking, evolutionary theories are *always* in the process of emerging. But in the early twentieth century, it is not just evolutionary theories that follow such protocol: scientific speculation about social development, at large, does so too. The scientific attempt to create objective truths about the formation of human society – and to institutionalize these truths – is enabled by a cultural narrative of national expansion.

Regarded as a precursor to the storyline for 'the current transnational hegemonic success of capitalism and the West', Henry Adams's *The Education of Henry Adams* (1907) fixates on discovering natural analogues for the capitalist machinery crucial for sustaining US global influence. According to John Carlos Rowe, Adams counsels (albeit indirectly) 'the most powerful leaders of his age', in particular his friend and former Secretary of State John Hay, 'to pursue policies that would consolidate the twentieth-century economic authority of the United States in international trade and serve as the [basis] for that special brand of US *neocolonialism* understood in the era of the Vietnam War in the phrase "establishing spheres of influence"'.[24] We will return to Adams a bit later; at present, I want to emphasize how the idea of establishing spheres of influence channels (neo)colonialist energies toward creating sites of international trade through which information flows. The rise of new academic disciplines, as we will now see, mirrors this process.

Anthropology is one such rising discipline whose emergence can be seen in the pages of popular science journalism. Reprinted in the

November 1895 edition of *Popular Science Monthly*, the address of the retiring president for the American Association for the Advancement of Science, Daniel G. Brinton, illustrates how new disciplines assert their claims to knowledge production. In his address, 'The aims of anthropology', he points out that the discipline is 'one of the newest of the sciences; in fact, it has scarcely yet gained admission as a science at all'.[25] Anthropology attempted to garner recognition by initially questioning evolution as a *final law*, asserting that in reality it remained 'a provisional and partial result of our observations'.[26] Anthropology then *colonized* this intellectual space, creating an alternative model, the search for universal cultural traits. This search, according to Brinton, constituted a 'careful study of what are called "ethnographic parallels" – that is, similarities or identities of laws, games, customs, myths, arts, etc., in primitive tribes located far asunder on the earth's surface'.[27] Cued by the reference to 'primitive', the 'ethnographic' mindset was guided by a colonialist logic that animated the standard litany of explanations for the inequality of races. Furthermore, Brinton's account of what anthropologists do seems intended to put their discipline in charge of establishing scientific truths. After asserting the importance of 'objective investigation', he states: 'The aim of the anthropologist in this practical field is to ascertain in all their details, such as religions, language, social life, notions of right and wrong, etc., wherein lie the idiosyncrasies of a given group, and frame its laws accordingly'.[28] It should be evident that Brinton chastises evolutionary science for doing the same thing he now recommends that anthropology should do: discover 'laws' extrapolated from objective investigation. Indeed, he envisions an entire restructuring of academic disciplines around anthropology: '[I]n the future, political economy, like history, will have to be rearranged on lines which this new science dictates'.[29]

Brinton strategically draws on the cultural capital of 'hard facts' and 'thick contexts' to support the abstraction of science. Making clear the distinction between physical characteristics and speculative generalities, while holding them to the same degree of validity, he insists that the concept of 'a racial mind, or the temperament of a people' can be discussed 'with as much propriety and accuracy as we can say of any of the physical traits which distinguish it [a race/people] from other people or races'.[30] At the same time, he announces, 'the final arbiter' is 'not the social group, but the individual'.[31] To make sense of this shift in criteria and how it serves a larger cultural purpose, we might turn to Mark Seltzer's study of 'statistical personhood' in *Bodies and Machines*. Like Brinton, Seltzer disavows 'an opposition or separation of abstract "forms" and material "practices"'. For Seltzer, the making of statistical persons relies upon 'the relays between the formal and the material, the abstract and the practical, the generic models and the concrete individuals'.[32]

That the search for ethnographic parallels, in anthropology, ultimately takes place at the level of the individual rather than the group, demonstrates the tensions between abstract and realistic conceptions of the self that Seltzer explores, tracing them through Jacob Riis's study of the social effects of immigration, *How the Other Half Lives* (1890). According to Seltzer, 'Riis's photographs routinely juxtapose rounded or shapeless human forms against strongly lined, almost gridded backgrounds'.[33] In his exposé of urban slums and call for social reform, Riis uses a 'generic model' to represent 'concrete individuals', an abstract figure standing in for a *type* of person.

Not only is this practice echoed by Brinton's oscillation between universal traits and individual identity; the action is also imitated by the idea of national character, an imagined sense of belonging through traits that all citizens appear to possess. National character is and is not reducible to the level of the individual; it meets and yet exceeds the category of the singular self. In this light, it is easier to grasp Brinton's racial mind as a property of both the individual and group.

But immigration, a conjoining factor with US expansion, threatens the idea of national belonging. In 'Race mixture and national character', published in *Popular Science Monthly* in May 1895, Lewis R. Harley remarks: 'The introduction of large numbers of foreigners into our country leads us to inquire whether there is such a thing as national character in the United States'.[34] Elaborated by Brinton, the anthropological strategy tried to hold together the fractured national self through the idea of a universal identity, one that both promotes and erases racial divisions. Brinton's proposal of a racial mind further reflects the ambivalence of biological markers; as Walter Benn Michaels reminds us, 'blood is blood, but – because blood is blood – blood isn't enough'.[35] A materialist approach only goes so far, a sentiment that underwrites both the abstractions of scientific knowledge production and the making of statistical persons. Referencing Herbert Spencer's thinking, Harley claims that '"the mixture of races" is a theory of evolution, and the unity that is developed is not of blood but of institutions'.[36]

FIGURE 1 'Mullen's Alley, New York', from Jacob Riis, *How the Other Half Lives*

It could be said that statistical data analysis – ranging from economics to climatology – contributes to a 'bloodless' solution to Harley's problem. Setting the stage for such a solution, Brinton insists that scientific observation is spelled out in the primacy of the field report. Acts of observation could be meshed together to provide a workable worldview. The reporting on overseas lands, in *Scientific American*, becomes coloured by statistical data analysis that reinforces the 'value' of global relationships to an emerging national power. This is a sign, I believe, that popular science journalism is appealing to a readership familiar with the premise that economic expansion beyond national borders is speaking a *similar language* in expressing a desire for control of a region's exports and imports and a compulsion, as Rowe reminds us, to assert US 'economic authority'.

'Our trade with Africa', published in the 20 June 1896 edition of *Scientific American*, starts with the report that the 'value of the [general] exports' to South Africa was set at the very least to double by the following June: from $5,000,000 to $10,000,000. While the article goes on to say that the US remains behind Germany (less than half) and England (less than a quarter) in total exports of manufactured goods to South Africa, 'American manufacturing plants are capable of turning out twice the amount of goods requisite for the supply of this country in a year'.[37] Here the statistical outlays project a future of US economic dominance.

But we are reminded that the US must vie with England, a nation with a similar trade philosophy. The account of Stanley's thoughts in 'The development of Africa' emphasizes how the continent has been carved up, with land ownership serving as the impetus for economic dominance. Yet England, like the US, appears to take a different approach by 'reserving spheres of influence'. Previously, however, England had been 'policing the African coasts for over half a century'.[38] To drive the point home that England has moderated its imperial desires, the article ends with a statistical explanation for the reconfiguration of the African region. In particular, we get a picture of a flexible strategy for developing 'nationalized markets', for England's holdings are sub-distributed across three commercial interests: the South African, British Central Africa and British East Africa companies.

Going beyond a purely economic purview, Gustave Michaud's 'The climate of our new possessions', published in the 15 September 1900 edition of *Scientific American*, relates the production of scientific knowledge to US global imperial adventures. Michaud announces that, due to US dominance in the Pacific, 'the scientific world is now getting its most complete and accurate information on many peculiarities of the tropical climate'.[39] Yet such knowledge is modelled around the climate experienced by the citizens of the modernized urban enclave of New York City. Statistical personhood becomes both globalized and naturalized; Michaud compares, through the use of a chart, the 'mean monthly temperature' of that city to the Pacific regions under

US ownership. Attempting to familiarize readers with these regions, he uses geopolitical centres of influence to trace a larger climate map: 'The climates of Manila, Honolulu, Havana, and San Juan represent probably fairly enough the relation existing between the climates of the Philippines, Hawaii, Cuba, and Puerto Rico.'[40] In setting the stage for US economic oversight, the Pacific regions are dissected by dint of their political lines of power and climate differences. We see here the connections between political and natural science that enable statistical language to convey imperial desires.

Statistics expressed as economic language, to be sure, would continue to inform US foreign policy. In noting the rapid increase of the Philippines' chief commercial product, sugar, M. W. Harrington's 'Natural products and resources of the Philippine Islands', published in the 4 June 1898 edition of *Scientific American*, shows the connections between scientific and economic reporting. Under the guise of objectivity, the physical landscape of the islands is broken down, restaging the way land was carved out or allocated by colonial interests. Remapping the landscape, imperial control itemizes (in economic parlance) the territory into commercial products: sugar, hemp, tobacco, cocoa and coffee, in order of their 'importance, as articles of export'.[41]

Statistical language then makes it possible both to detail the economic appeal of global trade and to justify the cultivation of spheres of influence through the apparent scientific objectivity of this language. Besides chronicling revenues, expenditures and exports, an article, 'The Philippine Islands', published in *Scientific American* in 7 May 1898, demonstrates how the reshaping of global relationships – expressed in the language of scientific objectivity – animates the exploitation of natural resources: 'There are large coal fields which are now being opened, so that it is expected that 5,000 tons of coal per month may be mined'.[42] We then see how technology and information constitute a feedback loop that connects US spheres of influence, a global project central to Hay's foreign policy: 'There are 70 miles of railway on the islands and 720 miles of telegraph'.[43] Another article, 'American trade with Central and South American countries', published in the 7 May 1898 edition of *Scientific American*, is even more explicit in connecting statistical personhood to global trade. The preface to a chart of population figures of each country featured in the discussion reminds us: 'We do not always realize the importance of our trade with our southern neighbors'.[44] While cultural difference is strategically reduced through the idea of creating geopolitical 'neighbors', the rhetorical engagement with readers is intended to reinforce the ideology of the 'expert' observer who, it is suggested, sees what the reader cannot see. Thus an imperialist worldview is coterminous with one that is defined as scientific.

Hay reinforces the objectives of popular science journalism in his view of 'foreign policy in terms of the commercial interests he was convinced would replace military conquest as the focus of twentieth-century international politics'.[45] It is crucial, moreover, to recognize that Hay's proposals, such as the Far Eastern 'Open Door Policy', rest 'not on territorial or even spatial domination', but rather 'upon the command of "representational" power either in commercial markets or their equivalents in cultural productions'.[46] Hence the reproduction of power through multiple cultural channels is complemented, as we have discussed, by the emerging value of observation in the scientific marketplace. Just as popular science journalism highlights the role of political, information and economic networks, US foreign policy does the same – and both engage with the dictates of modernity, that is, the fluid connections between these networks projected on a global scale. Rowe suggests that Hay '"originated" nothing, merely playing with the historical fragments he acquired . . . In the balance-of-power politics that he played, Hay composed a new authority for the United States that bears comparison with that of the modern literary author's'.[47]

Such a comparison is accurate for another reason: according to Luhmann, modernity raises the issue of artistic presentation, but does so by rewriting this issue as a matter of second-order observation. In the modern era, Luhmann explains, the making of art 'amounts to creating specific forms for an observation of observations . . . From this perspective, the artwork accomplishes the structural coupling between first- and second-order observations'.[48] Hay's 'playing with the historical fragments he acquired' suggests the same kind (if not degree) of 'structural coupling'. As a result, 'cultural productions', that is, popular science reporting, art or the 'command of "representational" power', all create the means for carrying out observations. The 'new authority' of the US depends upon the feeding back of free market ideology, constituted by 'commercial interests', through informational observing systems. Here, what Wolfe says about Luhmann's vision of modernity, that it legitimates 'the asymmetrical importance of the economic system',[49] resonates with Hay's model of foreign policy. In light of the parallels we glimpsed earlier between national expansion and the global reach of popular science journalism, Hay's rewriting of national authority through commercial interests is similar to the revision of the role of natural selection in evolutionary theories: the 'paradigm' for both, as suggested in Livingston's remark cited above, is 'displaced as it achieves greater *reach and saturation* of other realms; its pluralization and relativization make it more resilient even as they may in some ways make it less *itself*'.[50] For example, representational power circulates through the structural coupling of first- and second-order observations that 'displace' authority through political, information and economic networks.

Indeed, the mission statement of *Scientific American*, articulated in a 28 December 1895 editorial, echoes the transformation of national force – prevalent, for instance, in the magazine's habitual in-depth reporting on new and more powerful battleships – into economic coercion: a dedication to 'the arts of peace' guides the exploration of the 'world of science'.[51] Later, the editor suggests that the Monroe Doctrine should be understood not literally but rather as an abstract guiding principle that can be modified according to the demands of foreign policy objectives. Like Hay, Norman Angell, in *The Great Illusion* (1913), forecasts a policy shift away from the emphasis on military power towards economic determinism. That is, according to Angell, not only does evolution punish militaristic nations by forcing these nations to sacrifice their 'fittest' citizens during wartime; in Herbert Spencer's much-rehearsed formulation, it drives social progress as modernization, a process guided by the increasing dominance of global trade.[52] That Angell regards nations no longer as 'persons' but rather as racially pluralistic entities bears out Livingston's comment that evolutionary 'nature/culture hybrids', addressed by popular science journalism, relate to 'the implosions of diversity back to the imperialist metropole'.[53] Not only, then, do racial hierarchies appear to fade only to return in the cultural logic of imperialism; national self-image is reproduced in an emerging vision of the economics of power. While we may be tempted to call Angell's vision of economic cooperation post-national, in actuality the supposed relativization of national identity through global trade, as Livingston points out in a passage already cited, 'seems to have been one of the leading principles of empire building'.[54]

Global(ized) trade, moreover, contributes to Immanuel Wallerstein's analysis of a 'capitalist world-economy'. For Wallerstein, 'the imperative of the endless accumulation of capital had generated a need for constant technological change, a constant expansion of frontiers – geographical, psychological, intellectual, scientific'.[55] While such thinking fits remarkably well with Livingston's 'rough-and-ready account' of global capitalism, it all the more strengthens the ties between scientific observation and academic disciplinary formation. That is, how to understand this 'capitalist world-economy' as a *system* created an academic 'division of labour' that categorized disciplines, such as anthropology, according to their fit within a science or humanities framework. The historical divide between the sciences and humanities 'was premised on a certain structure of the *world-system*: a world dominated by the West, in which the "rest" were either colonies or semicolonies'.[56]

III. Information and control systems

At this juncture, we are ready to examine further the mechanisms of US authority in the global marketplace by making a theoretical intervention into the process by which systems of informational control are constituted. Let us begin with Seltzer's encapsulation of the development of these systems:

> The radical transformation in the thinking about programming and control from the 1870s to the 1930s that James Beniger has called 'the control revolution', involved, above all, a rethinking of the problem of representation, communication, and information-processing: that is, the understanding of processes of representation – the always material forms of information-processing – as production, and the understanding of production as processing, programming, and systematic communication.[57]

Such a revolution, Beniger explains, is guided by an economic model that regards 'society as a *processing system*, one that sustains itself by extracting matter and energy from the environment'.[58] Social systems, modelled after living systems in that they are open rather than closed to their environments, follow the second law of thermodynamics: they process material and energy flows to forestall dissipation. As Beniger puts it, 'control is essential to life itself';[59] communication is crucial to the ability of systems to maintain their organization. Accordingly, he quotes from Luhmann: 'The system of society consists of communications. There are no other elements, no further substance but communications'.[60] Emphasizing the importance of control, as Beniger does, raises a question: does the connection between control and communication support a turn to teleology, in that the processing of systems appears to be programmed? Beniger would answer that such a concern has already been addressed by biologists; as he explains, 'Programmed behavior is not teleological but *teleonomic* . . . In [Ernst] Mayr's words, "a teleonomic process or behavior is one that owes its goal directedness to the operation of a program . . . leading it toward a given end" '.[61] Such thinking was espoused by Spencer, who regarded society 'as an integrated system',[62] regulated by an evolutionary programme. As systems became linked to thermodynamics, James Clerk Maxwell's 'demon' that challenged the second law by turning entropy into recoverable energy, according to Beniger, insisted 'that control involves programming, that programs require inputs of information, that information does not exist independent of matter and energy and therefore must incur costs in terms of increased entropy'.[63]

While Beniger surmises that 'programs control by determining decisions', these decisions are not governed by choice or free will: 'All decisions are determined in the sense that, for a given program, the same input will always result in the same decision or output'.[64] But if the 'control revolution' to some degree departs from free market ideology, at the same time it espouses the colonizing logic of this ideology. This is what Livingston sees in the flexibility of evolutionary thinking about systems and what reappears in Beniger's argument: 'Information processing and programmed decision are the means by which such material processing is controlled in living systems, from macromolecules of DNA to the global economy'.[65] Connecting the thermodynamic language of 'material processing' to globalization, Beniger, at this point, contends that programming has yet to receive adequate attention, blaming cybernetics for overstating the importance of 'control behavior and feedback'.[66]

But is cybernetics to blame, as Beniger contends? Ever since the advent of modernity, the contingency of observation, which we will discuss in a moment, has hindered attempts to formalize a centralized programme that initiates the decision-making process. Contingency, furthermore, Wolfe posits, extends to 'all observations, including those of nature'.[67] Beniger believes that natural selection is crucial to programming, for he supports the idea that selection, operating at multiple levels (group, kin and reciprocity), guides social development and the evolution of control.[68] Yet programming cannot be granted value simply because it is affiliated with evolutionary principles, such as natural selection, for the model of evolution has been thermodynamically transformed through the emergence of the principle of self-organization that shapes the constitution of systems.

This precursor to or prehistory of cybernetic systems-based thinking, Erich Jantsch remarks, turns on a central principle: the idea that a self-organizing system '"knows" indeed what it has to import and export to maintain and renew itself. It needs nothing but the reference to itself'.[69] As a result, what is known as second-order systems theory views self-reference as critical to what Beniger would call programming and what Jantsch and Luhmann would call the system's self-making, or autopoiesis.

The take-off period of global systems is enabled, as Luhmann reminds us, by the condition of modernity that foregrounds questions about representation. In this light, Hay inaugurates a new mode of foreign policy, one that is 'intent on establishing a US imperium – not in lands, peoples, or even raw materials, but in the control of nations and, of course, their markets'.[70] In Hay's abstraction of national authority, all parts of the globe are connected by feedback loops. Thus he has to account for overlapping patterns and networks of communication and information at the national, economic and regional level. By creating and maintaining spheres of influence through these patterns and networks, Hay fashions an integrated programme for global management: one founded on informational control that extends from the materiality of production to encompass strategies of abstract representation, that is, information as a self-reflexive system that appears to operate beyond corporeal limits. Therefore information becomes regarded as a political commodity; the mapping and researching of overseas regions becomes a national imperative. It is an activity in which popular science journalism participates and legitimates through acts of observation.

But the turn into the twentieth century produced new types of scientific theorizing around genetics and radiation that reveal the limits of observation. Karl Pearson's *The Grammar of Science* (1900) marks the prevailing mood of uncertainty: 'In the chaos behind sensations, in the "beyond" of sense-impressions, we cannot infer necessity, order, or routine, for these are concepts formed by the mind of man on this side of sense-impressions'.[71] With predictable dismay, Adams summed up what Pearson had implied: 'In plain words, Chaos was the law of nature; Order was the dream of man'.[72] Granted, Pearson, unlike Adams, was able to accept if not fully embrace the constructive possibilities for complexity, which leads to the distinctions between first- and second-order observation. Adams, though, signals the dangers of letting anxieties about solipsism as such become assumptions about how systems and their relations with their environments are constituted. As Livingston reminds us, those anxieties can be internalized by systems themselves, '*positing a world of naturally solipsistic creatures*'.[73] His reading of second-order systems theory suggests that we regard the notion of solipsism as a 'false problem in the first place, leaving us to start again if we can'.[74]

The reason why solipsism is a 'false problem' can be grasped through Luhmann's writings. As already stated, Luhmann has formalized the contingencies of observing systems. Elaborated further, while every observer has a blind spot that can only be seen by another observer who is blind in the same way, paradoxically, it is this blindness that grants access to the world. According to Wolfe: 'What links the system to the world – what literally makes the world available to the system – is also what hides the world from the system, what makes it unavailable'.[75] The world does not exist before or beyond us, but through us. Another way to describe this condition would be to consider it as a performative effect.

This implies that identity markers such as race, class and gender, and scientific information, are co-constituted in the ongoing US imperial projects that reshape the globe and extend citizenship beyond national borders. In particular, as Barad would emphasize, self-reflexivity is critical in accounting for the contingent status of our observations, since 'humans do not merely assemble different apparatuses for satisfying particular knowledge projects; they themselves are part of the ongoing reconfiguring of the world'.[76] Not only is the globe being reshaped; race, class and gender categories as scientific data also come to be constituted through the second-order observations characteristic of global information systems under the regime of modernity. There is no longer a de facto raced, classed and gendered observer observing: along with the complexity produced through observing systems, scientific practices do not come before or after signifiers of (national) identity.

IV. Historicizing science

The proclivity of popular science journalism for second-order observation, coupled with Barad's revision of scientific practice, should finally put to rest the belief discussed in the introduction that evolutionary theories were a monolithic formulation fundamentally based on social Darwinism or neo-Darwinism. I want to focus on one historical case study that supports such revision: a reprint in *Popular Science Monthly*, April 1909, of an address, 'Life and works of Darwin', by Henry Fairfield Osborn delivered at Columbia University on the hundredth anniversary of Darwin's birth, the first of nine lectures on 'Charles Darwin and his influence on science'. Given the context, we might expect a non-critical form of scientific boosterism for Darwin's thinking. And yet what we do get, in fact, is a disciplinary critique that, at times, discloses the performativity of Darwin's theories. Osborn first points out that Darwin's '"variation" and "variability" are very broad and elastic terms. Every actual example he cites of his main hypothesis . . . is a variation both of heritable and of adaptive value'.[77] Already there can be no monolithic reading for Darwin, and going still further, Osborn challenges the validity of Darwin's central theoretical tenet: 'Darwin did love his selection theory, and somewhat overestimated its importance'. In particular, Osborn discusses three points of Darwin's thought that are in his view no longer accepted. First, pangenesis, overturned by '[August] Weismann's law that heredity lies in the continuity of a specific hereditary plasm, and for want of evidence of the transmission of acquired characters'. Second, the continuity of changes: 'Many changes are discontinuous or mutative, as held by [William] Bateson and [Hugo] De Vries'. And third, Darwin's 'belief that out of fortuitous or undirected variations in minute characters arise direction, purpose and adaptation through selection still lacks proof by either observation or experiment'.[78] Osborn historicizes Darwin, citing the current challenges to his theories and reminding us that Darwin's theories are products of a specific place and time. This limits any attempt to see these theories as universal and timeless, so often the endpoint of observation regarded as transparent and of first-order status.

Thus there was no consensus about evolutionary theories. Rather there seems to have been ongoing debate. My point is that such debate had no preordained outcome, and, even if it appears plausible that the outline of evolutionary theories remained intact while the content was revised – that is, what Livingston stated in the second section of this chapter – it seems even more feasible that the development of evolutionary thinking depended on the kinds of transformations that only self-reflexivity could make possible: glimpsed through the constructive possibilities for second-order observations staged through the co-constitution of the scientific and social realms.

The case study, moreover, has much to say about the gendering of scientific observation. Historicizing science (through the reading of science in the making), to put it directly, initiates a challenge, voiced by Livingston, to 'the coding of closure with science and masculinity and of openness with a feminized humanism/humanities'.[79] In the chapters that follow, I will read the formation of identity depicted by literary narratives against second-order observation to disclose key moments of national subjectivity seen as a performative category: one that puts gender, as well as race and class, back into play.

2

DIRTY NATURALISM AND THE REGIME OF THERMODYNAMIC SELF-ORGANIZATION

Through the perspective of popular science journalism, evolution becomes an artistic question. In a rather astute public relations coup, *Popular Science Monthly*, in 1895 and 1896, featured articles by Herbert Spencer on the development of various professional institutions. Taken collectively, these articles seem to follow a similar trajectory, confirming Spencer's view of evolution driving cultural progress. In particular, his study published in August 1895 of the development of the orator, poet, actor and dramatist legitimates what he sees as a tendency towards integration, diversification and complexity. In his vision of cultural progression, a racial hierarchy is legitimated as 'germs' of these emerging professions had 'evidence, supplied by uncivilized races and by early civilized races'.[1] Likewise, his discussion of painting, published in March 1896, traces the process of specialization, which lines up with his beliefs about the evolutionary idea of organization that broadly reflects a doctrine of self-improvement and social reform.[2]

The interest in the relationship of art to science continues as writers both challenge and expand upon Spencer's thinking. In John Quincy Adams's 'Science versus art-appreciation', published in *Popular Science Monthly* in March 1903, the relationship is initially antagonistic, as Adams takes a contrary, even derogatory, view of Spencer's picture of the evolutionary progress of culture:

> We see nearly every field of human effort controlled by specialization. It is no less clear that the aim of the manufacturer, as of the investigator, is to set forth 'the latest thing out'. Every province of human interest has been brought under scientific classification, so that nearly all

> thought is now cast in 'general ideas'. This mode of thinking ignores individuality and sees in men and things only units of a class.[3]

Adams sees art as not fitting into a schema based on levels of organization and complexity, for at heart, the creative impulse cannot be quantified, much less fitted into a scientific process. He insists that 'a dictionary of artistic language is impossible. Not a line, a form, a color nor a rhythm stands for a definite idea, the meaning of one depending not only upon its relative position as expressed by the artist, but also as appreciated by the spectator or listener'.[4]

The case that Adams makes relies on two critical ideas that constitute systems theory. First, his statement of self-organization, 'relative position', does not rely on a universal principle of organization and complexity – that which characterizes Spencer's writing about art – instead showing that the whole of the artistic work emerges greater than the sum of its parts. Second, what builds up internal complexity is the act of observing art. In framing observation, Adams focuses on aesthetic response, 'a direct appeal to one's own internal measure, called appreciation'.[5] At this point, a consensus is reached with Spencer, as Adams, likewise, argues that art has a social function, one that leads to cultural reform. In a gendered image of commerce and value exchange, a woman who knows the difference between a superior piece of furniture and an inferior one signifies the aesthetic judgement of art:

> If a woman, with a good aesthetic standard, goes forth to buy furniture, she is no longer in a mood to be persuaded to buy an object, unless it comes up to her conception of beauty. Neither gilded ugliness, expensive tawdriness, nor the 'latest thing out' is wanted, but a character which she can live with and enjoy. Such a demand on the part of a goodly portion of purchasers would materially change the character of our manufactured product and leaven our social and industrial life.[6]

While Adams begins his article opposed to Spencer, he ends up expressing an objective criterion for art, voiced in a 'standard' of cultural refinement that conceals class-based tastes in the same language of universal value used by Spencer.

Four years later, science is used to redefine art. Reacting against a class-based privileging of the fine arts, Ira W. Howerth's 'The classification of the arts', published in May 1907, suggests an 'evolutionary classification', based on 'the order of complexity and potential unity'.[7] Taking Spencer's paradigm further, he claims that

> the botanical and zoological arts, which are capable of bringing into existence new forms of life, ideal forms, differing in size, shape, color and character from anything that nature has produced, are also creative arts . . . There seems no absurdity, then, in ranking above the art that paints a flower the art that can produce one; above the art that beguiled the birds, the art that can change the leopard's spots.[8]

Howerth places fine arts at the bottom of a social hierarchy; the greatest complexity and unity is exhibited in the artful practices that seem closest to imitating evolutionary development, such as agriculture, education and the breeding, domestication and training of animals. At the top is what he, borrowing from sociologist Lester Ward, calls 'sociocracy', that which denotes how social reform can be seen as a meta-art, 'the art of arts', spanning all artistic endeavours to achieve a 'perfected humanity'.[9]

Theodore Dreiser's *Sister Carrie* (1900) and Edith Wharton's *The House of Mirth* (1905) connect the social function of art to a regime of widespread reform. Unlike the *Popular Science Monthly* writers, they challenge how cultural values are viewed as straightforward products of uniform evolutionary processes. Their repeated scenes of theatrical plays and art shows suggest that who is doing the observing matters as much if not more than what is being observed, circling back, through self-reflexive narrative strategies, to demonstrate the contingency of all observations of social and natural environments – and how fluid the designators for these environments are. In their depiction of a naturalist shift from survival of the fittest in natural environments to economic competition in urban environments, Dreiser and Wharton develop a different picture of psychic and social systems, one that casts light on the plight of the socially vulnerable and politically marginalized. This picture develops through a study of art as a social system and how observation is interrelated with social regulation. By pressuring observation (especially refracted through art), we can interrogate the associations between biological and economic decline. As I will illustrate, changing scientific paradigms (from biological evolution to thermodynamic self-organization) do not dislodge a prevailing ideology of acquisition – one that rapidly becomes globalized.

Such ideology, even now, envisions the future as an ideal(ized) outcome of scientific theories of self-organization that supposedly guarantee a moral equilibrium brought about by individuals working to be 'the best they can'. Dreiser's *Sister Carrie* and Wharton's *The House of Mirth* express an oppositional vision, rewriting the traditionally harmonious, open, clean and organic 'outside' of naturalism as 'inside' spaces of chaos, confinement, waste and pollution. Their narratives create what I call *dirty naturalism*, challenging the 'greenwashing' of structural economic inequalities through the downward mobility of George Hurstwood (in *Sister Carrie*) and Lily Bart (in *The House of Mirth*).

I. Social regulation and the power of art

Commenting on how *Sister Carrie* can be read as a narrative about both modernism and evolution, Jennifer L. Fleissner finds that we tend to regard the two storylines, Carrie's rise and Hurstwood's fall, as polar opposites: Carrie represents a crass consumerism informed by fantasy, while Hurstwood acts out a maladaptive evolutionary outcome in a realistic social setting. Fleissner inverts these readings, arguing that Hurstwood's story is depicted in exceedingly sentimental terms (he becomes a victimized 'old man' deserving of our pity) and Carrie's story, conversely, is a realistic study of the new woman not as a consumer but as a worker (a narrative that challenges our understanding of early-twentieth-century gender and labour relations).[10] Thus for Fleissner, Carrie's rise embodies a female agency constituted through work. Fleissner does not, however, go far enough in historicizing Dreiser's novel, for a lingering fantasy remains: the woman as a free agent who earns her independence. This problem is critical because the foregrounding of Carrie's story encourages us to overlook, in Hurstwood's decline, the structural economic inequalities that underwrite Carrie's free agency. These inequalities register, through a thermodynamic image of metabolic loss, on the bodies of the destitute class, 'all pale, flabby, sunken-eyed, hollow-chested',[11] to which Hurstwood belongs at the end of the novel.

Carrie's success is set against a modern era of increasing complexity, staged through her resistance to becoming 'reduced to the common level of observation',[12] that allows a more nuanced model of observation to emerge. There are initially the second-order observations constituted through the narrator's perspective – that is, the observing of other characters' observations, often commenting on what these characters cannot see. For example, there is the scene in which Hurstwood and Charles Drouet talk excitedly about Carrie, while at the same time unable to witness, as the narrator points out, 'when the little toiler was bemoaning her narrow life'.[13] From there, we arrive at characters observing other characters' observations: after Drouet glances at a pretty girl while walking with Carrie, he says to her, 'Fine stepper, wasn't she?' Her response is to think that 'if that was so fine, she must look at it more closely'.[14]

Granted, such sequences of observation may seem like the foundational structure of many literary narratives. But here Dreiser introduces an element of theatricality that will lead to a further disquisition on performativity and art. The narrator remarks on what has just taken place, suggesting that we can only understand the meaning of the scene through the observations themselves,

> When one of her [Carrie's] mind sees many things emphasized and reemphasized and admired, she gathers the logic of it and applies accordingly. Drouet was not shrewd enough to see that this was not tactful. He could not see that it would be better to make her feel that she was competing with herself, not others better than herself.[15]

Not only do we notice Drouet's clumsy conversational gambit, which will eventually cause him to lose out in the competition for Carrie; we also perceive her start as an actress when 'instinctively, she felt a desire to imitate' the woman upon whom Drouet had commented. The idea of theatricality is then foregrounded in a scene where Carrie and her two suitors, Drouet and Hurstwood, watch a play, *The Covenant*. Afterwards, Drouet and Hurstwood remark on a scene in which 'the wife listened to the seductive voice of a lover in the absence of her husband'.[16] Drouet's lack of compassion for the husband ('served him right') reflects his unawareness that he has lost Carrie's favour. That the drama being observed ends up commenting, through irony, on its being observed demonstrates how plays, categorized as art, have 'something like a privileged critical relationship to society', as Cary Wolfe explains, because 'they literally fix social space'.[17] Systems theorist Niklas Luhmann, in *Art as a Social System*, further remarks on artistic works that 'their significance as objects implies a realm of social regulation', a realm most clearly delineated in 'the more complex case where one observes other observers by focusing on the same object'.[18]

As we will now see, 'social regulation' intersects with identity formation. When Drouet encourages Carrie to take a part in *Under the Gaslight*, a melodramatic play put on by his local lodge, the narrator tells us that she is already 'possessed of that sympathetic, impressionable nature' of an actress. Yet in a sense, we suspect that Carrie is performing different roles that go, of course, beyond the theatrical stage, an idea emphasized by her 'passivity' and 'innate taste for imitation'.[19] Thus Carrie's identity is always being reconstructed. Furthermore, her big break as an actress in New York occurs when the stage manager observes her frowning at the 'feeling that she [had been] effectively shelved' during a rehearsal of a 'summer play'. Seeing the potential for an unexpected, thus memorable, comic effect, the stage manager asks Carrie to frown throughout the play. During the performance on opening night, her acting unhappy, accentuated by her feeling 'as if there were nothing to her part',[20] produces the intended effect, which is to elicit laughter from the audience. Here, Luhmann's idea of social regulation, earlier encompassed by Drouet and Hurstwood's desire for Carrie while she is acting a role during *Under the Gaslight*, is amplified and expanded: 'All the gentlemen yearned toward her'.[21]

That the scene on the New York stage tests the limits of intention(ality) is complemented by how Carrie's performance both reiterates social heterosexual norms and calls into question the idea of an authentic 'self' doing the performing. That is, according to Judith Butler: '[P]erformativity is thus not a singular "act", for it is always a reiteration of a norm or set of norms', while at the same time, 'reiterations are never simply replicas of the same'.[22] In other words, there are repeated references to Carrie's 'imitating' someone else, not being her 'self', but each 'reiteration' occurs in a different context shaped by a system of social regulation. So Carrie suggests that the idea of authenticity is created through a performance – which goes beyond the space of the theatre – guided by what others want and how others want to see her. Carrie's performance resonates with Butler's comment that 'the impossibility of a full recognition, that is, of ever fully inhabiting the name by which one's social identity is inaugurated and mobilized, implies the instability and incompleteness of subject-formation'.[23] What begins as a staging of the narrator's prior comment about art's 'desire to reproduce life'[24] culminates in the notion, as Wolfe points out, that 'art can be said to imitate nature only because nature isn't nature . . . which is simply saying that all observations, including those of nature, are contingent *and* of necessity blind to their own contingency'.[25] In other words, contingency trumps the question of what Carrie means, or intends, to do.

At key moments, the media intercedes to assist Carrie's rise, suggesting how such contingency (of an identity constituted through observation) can be socially regulated. Drouet is reminded by an announcement he reads in the *Evening News* that he needs to ask Carrie to act in his lodge's play.[26] The response to her performance on opening night is underlined by 'the critic of the "Evening World," seeking as usual to establish a catch phrase which should "go" with the town, wound up by advising: "If you wish to be merry, see Carrie frown"'.[27] Lastly, in the depiction of Carrie's social ascendancy, she becomes famous as an actress in New York through the interconnections between the stage star system and the media. Borrowing from Luhmann's phrasing, these systems are functionally differentiated: to handle environmental complexity, autopoietic (that is, self-making) systems differentiate themselves into subsystems, reintroducing the system/environment distinction as systems become environments for their subsystems. The stage star system and the media, in the novel, are separate and yet communicate with one another through the language of celebrity, most visibly represented in Carrie's name 'blazing in incandescent fire' on a neon sign at the theatre.[28] Her name, having become a public commodity, returns us to the earlier moment when her name was 'bandied about' through the conversation of Drouet and Hurstwood,[29] a doubling back, or reiteration, of communication.

The socially regulative, as Luhmann defines it, is connected to the power of art in Dreiser's novel. A reviewer of the New York play states that Carrie's dramatic expression, her frown leading to her popular success, illustrates that 'the vagaries of fortune are indeed curious'.[30] Later, Bob Ames, a scientist who plays the role of social commentator, says that 'nature . . . makes the face representative of all desire. That's what has happened in your case'.[31] He argues here that Carrie should move from comedy to more serious forms of drama. Thus Carrie has cultural capital because she can represent something difficult, if not impossible, to represent completely. But art as a social system (*pace* Luhmann's book title) appears localized and contained in the repeated figure of Carrie endlessly rocking. It results in a kind of atomized subjectivity; a void filled, as Walter Benn Michaels would put it, by Carrie's endless desire, signified by her moving up the cultural ranking of less serious (lower) to more serious (higher) genres of drama (the categorization of these genres is, of course, shaped by social regulation).[32]

Only desire, as Michaels's reading of *Sister Carrie* would have it, could resolve the issue of desire (read as the desire for self-maintenance) in art as a system of representation, which like all complex systems is autopoietic.[33] But for Luhmann, while a system is self-referential, it is not self-identical per se, nor are there 'point for point correlations between the system and environment',[34] because of the paradox of the internal code that constitutes the system. Using the example of the law to illustrate Luhmann's paradox, Wolfe observes that

> the first-order distinction between legal and illegal in the legal system is itself a product of the code's own self-reference – that is to say, the problem is that both sides of the distinction are instantiated by one side of the distinction (namely the legal – hence the tautology: legal is legal). But the tautological unity of this distinction may be disclosed only by a second-order observer, operating within another system and another code, which must remain blind to its paradoxical distinction if it is to use that distinction to process events for the system's autopoiesis, and so on.[35]

Regardless, Carrie has a subjective capital that transforms desire within the context of a larger global system of imperialism: her manager asks her, 'How would you like to try subduing London?'[36] In its imaging of power, such a question sets the stage for redefining self-making systems as self-organizing energy and economic flows: what thus comes into view is the emergence of thermodynamic economies, a concept that moves us away from regarding evolutionary theories as a primary guiding principle for collective and individual self-development.

The House of Mirth is also often read as an evolutionary parable: after being expelled from an upper-class environment by the scheming of Bertha Dorset, her more financially and socially powerful rival Lily Bart is unable to adapt to life in the working class. But should we be suspicious here of the use of scientific theories to explain social customs and practices – and especially, as the novel insists, those customs and practices so given to the whims of fashion? It is, I think, interesting that within the ideological framing of this reading is a rather stark image of economic exploitation, one that reprises arguments about evolutionary 'fitness' versus economic 'fortune'. After Lily's social 'fall', one of her still-loyal friends, Carry Fisher, and Lily's cousin, Gerty Farish, arrange for her to work in a hat-making workshop. While it is stated that Lily's fellow women workers are 'fairly well clothed and well paid' in a 'fashionable' enterprise, the narrator reminds us that they work to produce 'ever-varied settings for the face of fortunate womanhood'. These working women are, to be sure, not 'fortunate': they are stuck in a routine of 'sedentary toil' in a dreary environment where 'the youngest among them was as dull and colourless as the middle-aged'.[37] A few pages later, we overhear the women talking about the buyers of the hats, Bertha and her social circle, the same cabal who supervised Lily's expulsion. Bertha and her circle are socially powerful and influential, possessing a cultural capital that extends beyond their class, as such assets trickle down to 'this underworld of toilers who lived on their vanity and self-indulgence'.[38]

As we therefore notice in *The House of Mirth*, social inequalities become more troubling when we start to look at Lily's situation as symptomatic of an economic failure rather than an evolutionary one. Wharton signals this shift in perspective through Carry, who finds Lily 'social' work. Namely, Lily is hired because of her insider knowledge of the cultural elite to instruct the nouveau riche on the proper tastes and conventions to adapt. The narrator remarks: 'It was, in fact, characteristic of Carry that, while she actively gleaned her own stores from the fields of affluence, her real sympathies were on the other side – with the unlucky, the unpopular, the unsuccessful, with all her hungry *fellow-toilers* in the shorn stubble of success'.[39] Dale M. Bauer, meanwhile, observes: 'None of Carry's objects of compassion are sympathetic because of their spiritual or ethical qualities; their appeal lies in their social failures and economic deficiencies'.[40] At this point, an alternative way of viewing ethics opens up, one incompatible with Lily's insistence on holding onto race, gender and class-based values that direct the novel's ending: Elizabeth Ammons explains that the logic of these values dictates that it is better for Lily to die alone than to have her 'intrinsic' purity threatened by a relationship with Simon Rosedale (the only character in the novel who seems genuinely to appreciate and goes the furthest to rescue Lily), a Jewish character depicted with all the anti-Semitic loathing the upper class (and Wharton herself) could muster.[41] In this opposing definition that initiates a systems-theoretical reading of the novel, morality resides in a world of 'fluctuating ethical estimates', summed up by Lily's outlook: 'Moral complications existed for her only in the environment that had produced them'.[42]

This outlook, in many ways shared by Dreiser, appears as part of Wharton's larger response to modernity, which as Nancy Bentley points out is a profoundly ambivalent one.[43] One source of her ambivalence is perhaps the way in which modernity redefines ethics. Exerting a powerful influence in the novel, Wall Street, in the language of systems theory, can be regarded as a signifier of the modern interpenetration of economic, legal and information systems that are autopoietic. Reading Luhmann, Wolfe focuses upon his idea of functionally differentiated social systems that denies 'some untainted, unmediated space of subjunctive interiority . . . some unquestioned and unquestionable yardstick by which the socially and linguistically constructed code of ethics may be judged'.[44] In other words, ethics can only be observed through a specific system constituted through its own operating codes, and thus 'the distinction between morality and merely ethical "roles" is lost'.[45] Thus ethics become performative, defined only through a constant state of becoming.

That the idea of performance is prominent in Wharton's novel demonstrates how a performative model of ethics troubles the rigid dependence on a moralistic social order. Ammons identifies the *tableau vivant* scene (Lily poses as Mrs Lloyd in the iconic painting by Joshua Reynolds) as a staging of Lily's 'dazzling, overdetermined whiteness', setting up the visible stage management that occurs when Bertha expels Lily from her yacht on the French Riviera, a move that exposes the ethical limits of Bertha and her co-conspirators.[46] Not only does Wharton use repeated references to performance; she couples these images with allusions to observers observing. The person who observes the 'sublime stage-setting forgotten in a hurried shifting of scenes' is none other than Lawrence Selden, who experiences 'the renewed zest of spectatorship that is the solace of those who take an objective interest in life'.[47] A self-referential friction is generated when the supposedly principled world of the upper-class elite is self-consciously contrasted with the world of the nouveau riche. The 'social out-skirt' world of the Gormers, whom Lily is advising, strikes her 'as only a flamboyant copy of her own world, a caricature approximating the real thing as the "society play" approaches the manners of the drawing-room'.[48] Later, as Lily's social descent (engineered largely by Bertha) continues, she advises Mrs Hatch, an outsider multiply divorced 'from the West' whose 'Oriental indolence' associates her with Rosedale, both characters judged as social undesirables.[49] Here, the narrator remarks,

> Lily had an odd sense of being behind the social tapestry, on the side where the threads were knotted and the loose ends hung . . .

> Compared with [this] vast gilded void, the life of Lily's former friends seemed packed with ordered activities. Even the most irresponsible pretty woman of her acquaintance had her inherited obligations, her conventional benevolences, her share in the working of the great civic machine.[50]

That these social distinctions are being made by Lily figures her too as an observer, marking a change in her role-playing. Selden, from whose perspective Lily's expulsion is largely addressed, initially sees that performance, set up by Bertha, as 'the effect of a closing tableau'.[51] His words remind us of the earlier *tableau vivant* in which Lily had served as a spectacle. Perhaps the most prominent instance of Lily's social 'exposure', the *tableau vivant* both relieves for a moment and reminds us of Lily's vulnerability (to being exposed). This notion of exposure is further developed in the yacht scene where performance exposes heteronormative constraints on the body, for as Butler would aver, Lily's agency, her ability to perform, 'is to be found, paradoxically, in the possibilities opened up in and by that constrained appropriation of the regulatory law, by the materialization of that law, the compulsory appropriation and identification with those normative demands'.[52] The exposure of the body, moreover, is staged through images of observation as an embodied act ('the ugliness of things rasped the eye'), other bodily gestures ('The Brys . . . already thirsted for new kingdoms') and (in)digestion ('chronic dyspepsia').[53] Lily's moment in the spotlight is then framed by signifiers of a system of social regulation concerned with bodily conduct and indicates how the *tableau* scene is informed by this system, to the degree that Lily's observation about Mrs Hatch's lack of 'inherited obligations' appears rather ironic, for it is Lily who is trapped in a social 'machine' that appears unmistakably destructive (and that will be critically examined in Chapter Four's reading of *The Custom of the Country*). Carry, during the yacht scene, voices this idea twice in two pages. First, she warns Lily about the media presence around their group that helps to legitimate Bertha's manufactured scandal: '[T]hat horrid little Dabham who does "Society Notes from the Riviera" . . . he's telling everybody that you and [George] Dorset came back alone after midnight'.[54] Second, she creates a cannibalistic image of class relations, the nouveau riche as predators: 'There's Louisa . . . we're on the best of terms externally; we're lunching together; but at heart it's *me* she's lunching on'.[55] Taken together, Fisher's comments about a social competition stoked by economic wealth disclose the connection between second-order observation masquerading as first-order reporting and what Wolfe calls 'a certain consumerist schematization of visuality'.[56] Consumerism and its affiliations with observation, furthermore, support a campaign of global imperialism. Representing the boundless flow of capital, the powerful and influential Dorsets, Ammons reminds us, 'never question their right to sweep into any corner of the world, impose their notions of culture and civilization, buy off the locals, and set up temporary or permanent residence'.[57]

Social regulation, as an arbiter of Lily's fate, foreshadows the *tableau vivant* scene. We witness Mrs Peniston's dismay at hearing of the rumours about Lily's conduct conveyed by Grace Stepney, who has ulterior motives, using these rumours to suggest that she rather than Lily is worthy of inheriting Mrs Peniston's money. Then we observe the rumours about Lily 'treading a devious way'[58] unfold. She seems not to be concerned about her 'intimacy' with Bertha's husband – a friendship, in her view, that might mediate her conflict with Bertha – but she is more troubled by her relationship with Gus Trenor, who has a financial arrangement with Lily whereby he speculates (or so she thinks) on the stock market on her behalf. Lily grows increasingly aware of Trenor's interest in her beyond financial matters, an interest that she tries unsuccessfully to deflect by engaging him in

> sentimental small-talk which doubles upon itself like the paths in a maze. He was first puzzled and then irritated to find himself always led back to the same starting-point, and Lily felt that she was gradually losing control of the situation.[59]

Adding to his ire, his 'unmanageable mood', he has been heavily affected by the decline of the stock market.[60] After a less than successful weekend at Bellomont where the Trenors reside, with the atmosphere clouded by Judy Trenor's suspicions, Lily takes part in the Brys' 'general entertainment', given for the purpose of their social advancement – or, as put rather bluntly by the narrator, '[t]o attack society'.[61] Aided by Carry and directed by an esteemed portrait painter, Paul Morpeth, the scene comprises Lily's *tableau vivant* as the main attraction. The doubled sense of performance, the stage built within the staged/artificial world of the Brys (like the Gormers, they belong to the nouveau riche class), intensifies Selden's reaction as observer. Gerty also functions as an observer, reminding us of how the different strata of society are connected: Lily has been helping to solicit donations from the well-to-do audience for the Girls' Club, an organization which Lily has visited twice.[62] The staging of Lily's body brings together the diverse social spaces in the novel, reinforcing Luhmann's idea about social regulation constituted through observers focusing on a single work of art – an idea discussed earlier in my reading of *Sister Carrie.*

In a systems-theoretical reading of Wharton's narrative, it is critical to note the presence of second-order observation that informs the *tableau vivant* scene, which pulls us further away from the notion that observation is somehow 'natural', a notion that has throughout legitimated Bertha's campaign against Lily. To introduce Lily's performance, the narrator explains that *tableaux vivants* 'may give magic glimpses of the boundary world between fact and imagination',[63] and yet seeing Lily as Reynolds's 'Mrs Lloyd' makes it apparent that the 'tribute' is not to the art itself (the painting on which the image is modelled), but 'to the flesh and blood loveliness' of Lily, where

> [s]he had shown her artistic intelligence in selecting a type so like her own that she could embody the person represented without ceasing to be herself. It was as though she had stepped, not out of, but into, Reynolds's canvas, banishing the phantom of his dead beauty by the beams of her living grace.'[64]

Although this is a scene of life imitating art, Lily's 'living grace' also returns us to Wolfe's comment that 'all observations, including

FIGURE 2 A typical *tableau vivant*: photograph taken in the Dutch East Indies, 1898

those of nature, are contingent *and* of necessity blind to their own contingency'.[65]

Given the somewhat confusing sense of what we are looking at, overdetermined, we might say, by the limits of Selden's own subjectivity as observer, the logic of the scene for Ammons depends on its unmistakable racial overtones, especially the notion of 'type'.[66] I have been making the case for shifting the focus to how the scene gains significant meaning by being regarded as an ongoing act of observation. This is already hinted at by the notion that the 'picture' had no 'distracted accessories of dress or surroundings', reminding us that we are seeing Lily front and centre, as the background sets off 'the long dryad-like curves that swept upward from her poised foot to her lifted arm'.[67] From Selden's perspective, it is an image of presence: the 'poetry of beauty' that he can feel only when he is with her. Ned Van Alstyne remarks that 'there isn't a break in the lines anywhere, and I suppose she wanted us to know it!'[68] While the obvious philistinism of Van Alstyne spoils Selden's aesthetic appreciation of the 'real' Lily, an irony that hangs over the scene, his remark discloses the duality of observation, which Selden denies. That is, there *is* a 'break' in the lines, made by the observation itself and the realization that it is an observation, self-referenced by Lily's intention or desire to make it so. Thus her performance communicates this disclosure (or exposure) to the viewers, reinforcing Wolfe's comments about how observations are 'blind to their own contingency'.

The curtain falls, but the scene repeats itself. Lily, sensing her 'recovered power', has a second exposure (or, perhaps, overexposure), 'a second opportunity of showing herself to advantage', for she waits until the crowd disperses to rejoin them in the drawing room. After 'the circulation became general',[69] the chapter ends with Selden, who during the scene has been unsuccessfully pursuing Lily, caught up in a dialogue between Van Alstyne and Gus Trenor. Van Alstyne comments on the artifice of women contrasted with Lily: '[W]hat's a woman want with jewels when she's got herself to show? The trouble is that all these fal-bals they wear cover up their figures when they've got 'em. I never knew till tonight what an outline Lily has'.[70] Communication, as a system, repeatedly in the novel 'doubles upon itself', like the iteration of the *tableau*. It calls attention, as Wolfe comments, to the fact that 'the point of the system is to reproduce itself'.[71] And as we have noted throughout this scene, such a compulsion to repeat is, for Butler, evidence that '[P]erformativity is thus not a singular "act", for it is always a reiteration of a norm or set of norms, and to the extent that it acquires an act-like status in the present, it conceals or dissimulates the conventions of which it is a repetition'.[72]

II. Self-organization and energy flows

Under the sign of modernity, self-making systems foreground the idea of self-reference. According to Bruce Clarke and Mark B. N. Hansen: 'Systems' observations of their environment are internally and autonomously constructed by their own ongoing self-productions. In other words, to maintain their autopoiesis, (self-referential) systems must remain operationally (or organizationally) closed to information from the environment'.[73] And just as in Wharton's novel a values-based morality competes with a world of fluid ethical estimates, here self-organization – defined by physicist Lee Smolin as a theory that traces the emergence of life out of energy flows and material cycles[74] – competes with self-reference (marking the disciplinary distinctions between physical and life sciences). Yet for the moment, at least, that self-organization has become *the* far-reaching means of explaining the operations of systems suggests the popularity of thermodynamic principles over autopoietic principles.[75] As I stated at the outset, if reading capitalist values from scientific theories shapes the interpretation of *The House of Mirth* as evolutionary parable, then the regime of thermodynamic self-organization is similarly structured. Ira Livingston argues that self-organization is appealing precisely because it mirrors capitalist modes of flexible accumulation: 'We come to *recognize* self-organization because we have *attained* it to some critical degree. The common term for this process is simply *globalization*'.[76]

Through feedback loops, autopoietic systems become self-referencing, which highlights the centrality of information to thermodynamic models of self-organization. Erich Jantsch remarks that 'a dissipative [i.e. thermodynamic] structure "knows" indeed what it has to import and export to maintain and renew itself. It needs nothing else but reference to itself'.[77] That the relationship between thermodynamics and information has been traditionally expressed in terms of entropy – an idea introduced by Claude Shannon and Warren Weaver in 1949 – is, however, misleading. In statistical thermodynamics, Jeffrey S. Wicken explains, entropy refers not to disorder but rather to indeterminability; he thus proposes to replace the concept of entropy in information theory with that of complexity.[78] To elaborate, Luhmann states that 'the system uses its boundaries to screen itself off from environmental influences and produces only very selective interconnections'.[79] Not only can information not be directly received from the environment; in the process of handling complexity, complex systems such as societies (as we discussed earlier) reintroduce the system/environment distinction as the means by which they become functionally differentiated.

Regarding entropy, in particular, as increasing complexity leads to speculation about the emergence of self-organizing systems that operate far from equilibrium. Smolin proposes that life is 'a particular type of process which has emerged on top of the flows of energy and cycles of materials that characterize . . . open systems. Life perhaps might be seen to have evolved a way to ride these flows and cycles the way a surfer rides the flow of energy in water waves'.[80] But it is no mere coincidence that Smolin's surfer image is located in a scientific landscape animated by corporate business-speak; Livingston finds lurking in accounts of self-organizing systems 'bionomic tropes' that conflate economic and energy flows, for instance the idea that 'networking' life forms, over time, create a 'global co-prosperity sphere'.[81] Thus the move from life as a closed system, where entropy becomes destructive, to an open system, where thermodynamic processes promote complexity, does nothing to dislodge an underlying capitalist ideology. As Livingston sarcastically puts it: 'Life, the tragic hero of the late nineteenth century, becomes at the dawn of a new millennium a surfing CEO with a cell phone'.[82] If we are not exactly sure where such logic leads, Stuart Kauffman, an influential theoretical biologist, sums it up for us: '[T]he entire system appears magically to evolve to a poised state where, on average, each does best as can be expected'.[83]

Sister Carrie and *The House of Mirth* construct narratives that act out how the process of gradient reduction, for example, the de-differentiation of temperature or pressure, directs systems. Thermodynamic equilibrium becomes death, a sign of insufficient complexity that results in system breakdown. In the novels, the pairings of Carrie with Hurstwood and Rosedale with Lily replicate how a thermodynamic economy creates winners and losers – and illustrate the mixed results of victory. The losers are gradually cut off from the economic flows which sustain their energy levels, their downfall connecting economics to metabolic processes. That is, from a thermodynamic perspective, dissipation *rewrites* the familiar naturalist plot of decline, for such a decline is a failure to 'keep riding' the waves. And what that looks like in a world of structured economic inequalities that drive social systems is not surfing, but *sinking*. I want to argue then that images of sinking inform Dreiser and Wharton's critical assault on an emerging 'new economy' of information, speculation and globalization.[84]

While the repeated motif of Carrie rocking in her chair simulates the motions of surfing, her discussion with Ames, as she nears the peak of her success, reminds us of the corrupting influence of the free market. Ames, who has 'opened a laboratory' in the city,[85] appears to be the voice of conscience in a society dominated by the search for superficial pleasures; he advises Carrie to aspire to be a more serious dramatic actress. And yet, in reality, Ames – in the role of a scientific spokesperson for a thermodynamic economy – is really urging Carrie to do a better job of covering up for the fact that her stardom reveals how capitalist expansion feeds off self-organization and complexity. To exercise a greater commitment to the power of art, as Ames dubiously counsels, reflects how, according to Livingston, 'a relentless aestheticization of the apparently kinder and gentler paradigms of chaos and complexity keeps the ecology of violence that it describes hidden in plain sight'.[86] And such violence, in *Sister Carrie*, is aptly illustrated by Hurstwood 'sinking' out of sight in two scenes. That New York, when compared to Chicago, Hurstwood's former residence, is a more complex and functionally differentiated society appears to rationalize how free-market policies support an increasingly destructive competition: 'In New York, the roads were any of a half-hundred, and each had been diligently pursued by hundreds, so that celebrities were numerous. The sea was already full of whales. A common fish must needs disappear wholly from view – remain unseen. In other words, Hurstwood was nothing'.[87] Later, the destitute class, which Hurstwood has joined physically, exhibits the signs of thermodynamic degradation: 'Miserable food, ill-timed and greedily eaten, had played havoc with bone and muscle.' His individuality is eroded as he is absorbed into a 'class which simply floats and drifts, every wave of people washing up one, as breakers do driftwood upon a stormy shore'.[88] This striking image of the inverse of Carrie's surfing in the thermodynamic economy perhaps suggests, even, they *are* the 'wave' upon which she rides to stardom.

In *The House of Mirth*, Rosedale's understanding of the connections between markets predicts the flexible accumulation strategies produced by globalization. That he triumphs despite a virulent anti-Semitism might sound a faint note of social optimism in Wharton's narrative. But Rosedale's success brings to mind the infamous United

Colors of Benetton advertisement campaign that, on pretences of promoting multiculturalism, imagines the victory of money over racial division, while overlooking how international capital exploits local resources and labour.[89] To support my claim, we may turn to Lily's social 'crash' on George and Bertha Dorset's yacht. This scene implies that US economic expansion (which shapes Rosedale's success) has instilled social competition at the heart of global networks: in order to stay afloat, after being caught in an act of adultery, Bertha Dorset tosses Lily overboard with her counter-accusation that Lily has been seeing Bertha's husband. Like the unfortunate Hurstwood, Lily gradually sinks from sight.[90]

Perhaps the most troubling aspect of the capitalist logic lurking within self-organizing complexity, Livingston argues, is that, while this logic purports to redefine identity as a contingent process produced through global networks, what tends to occur is a narrative recovery of 'individualist selfhood and progressivist development'.[91] Indeed, both novels set against a backdrop of self-organizing systems stories about the individual self. Carrie's triumph as an actress, as Ames notices, demonstrates that her facial expression is a 'thing the world likes to see, because it's a natural expression of its longing'.[92] In other words, Carrie feeds back her audience's own dissatisfaction (which, we might surmise, has much to do with economic inequality) in a self-reflexive loop that magnifies her status as a celebrity. In a sense, she is, as the narrator proclaims, 'capital',[93] able to combine flexible strategies of improvisation and self-determination that displace her identity into media networks. Following Livingston, such a displacement makes Carrie's rags-to-riches celebrity status 'more resilient' as she becomes ever less 'herself', a staging of the 'rough-and-ready account of the current transnational hegemonic success of capitalism and the West'.[94] That Carrie's success often appears at times rather inexplicable – a result, it could even been said, of sheer luck – reinforces the view that the 'random behavior' of a self-organizing system somehow emerges into an optimum condition of stability. One look at the fate of Hurstwood, however, should make us think twice about Kauffman's scientific reformulation of the time-worn adage about living 'in the best of all possible worlds'.

Likewise, Lily's failure allows us a sentimental celebration of her self-sacrifice throughout the narrative (for example, paying back Trenor's loan that she was tricked into accepting, refusing to blackmail Bertha Dorset with the letters of Selden, one of Dorset's lovers) which obscures the larger systemic interdependencies (between Wall

Street speculation and the embedded gender hierarchies of the upper class) responsible for her fatal end.[95] Furthermore, in the earlier *tableau vivant* scene, Lily re-enacts the portrait of Mrs Lloyd, who as Michaels notes plays the role of writer, carving her name into a tree.[96] This gesture of self-reflexivity offers, in place of a collective resistance to a self-organizing free market economy, the image of the solitary writer. Indeed, Michaels remarks that the more Lily resists, the greater her market value becomes to social 'collectors' such as Selden.

Yet Livingston sees the literary narrative of thermodynamic self-organization developed not in naturalism, but in cyberpunk. Working out the possibilities of techno-capitalism in a stylish retro-future, novels like William Gibson's *Neuromancer* 'whitewash' the consequences of global economic expansion by distancing the neo-social Darwinist struggle of the lower class to survive in the grimmest conditions of the *real* world from the adventures of the computer hacker, a modern-day cowboy jacked into the *virtual* frontier. Livingston's critique of Gibson's cyberpunk narrative is anticipated, as I have been arguing, by the logic of self-organization visible in Dreiser and Wharton's novels. Fleissner remarks of the former, 'Carrie's body has been conceived . . . as cyborglike in its ability to replicate itself over and over in a new form',[97] which embodies a larger emerging narrative of global capitalism. And unlike Gibson, both Dreiser and Wharton challenge class separation. Most prominently, this challenge is articulated in scenes where Lily toils with the other female workers in the hat-making shop and the downtrodden Hurstwood encounters the successful Carrie, exemplifying the return of what a capitalist-guided morality tries to repress: the wasting of human lives through dissipation. As we will now see, thermodynamic economies disclose the waste generated by Wall Street speculation in a way that can serve as an ecological warning about overstressed systems.

III. Ecocriticism and thermodynamics

Wicken's earlier proposal to evaluate information according to its complexity seems apt with regard to recognizing ecological problems (which are, to be sure, highly complex). And yet it seems that responses to such problems are invariably slow. Hence it is necessary to look at how autopoietic systems are 'coupled' with their environment. Luhmann's concept of functional differentiation impinges on environmental issues, because 'to explore how a society can react to the exposure of ecological dangers the constraints on the possibilities of its subsystems must be examined'.[98]

The primary function for systems is to continue their autopoiesis. Although any concerns for the environment must come after a system's autopoietic maintenance, from an evolutionary perspective, as Luhmann comments, 'systems pursuing a trend toward exposure to ecological self-endangerment are eliminated'.[99] Viewed within an evolutionary framework, moreover, autopoietic (living) systems are thermodynamically compelled, that is, they depend on the energy-processing capabilities of the ecosystems that sustain them. It is here, then, that we can make an ecocritical intervention, refashioning Luhmann's system/environment distinction into an evolutionary-shaped model of ecosystems that operates according to thermodynamic principles.

The thermodynamic process of gradient reduction is at the forefront of this intervention. As Eric D. Schneider and Dorion Sagan explain, nature's abhorrence of gradients 'is an unstoppable tendency where energy flow leads to different natural complex systems including life itself'.[100] Consequently life can be defined as an open (not closed) system that operates far from equilibrium. Such a system exchanges an internal order for an external order, exhibited as forms of waste, like pollution. Because efficiently managing energy flows is necessary for a system to maintain its autopoiesis, Schneider and Sagan regard natural selection as making choices at a 'higher' level. Viewed as important units of selection, energy flows of ecosystems trace out the larger direction of evolution, 'the equilibrium-seeking organizations of open systems thermodynamics'.[101] A thermodynamically shaped model of selection is 'more ecological' and redefines natural selection, according to Schneider and Sagan, as 'how organisms, species, and populations mesh with their encompassing ecosystems and environment'.[102] So if we are somewhat slow to pick up the ecological message, as Luhmann suggests, we are nevertheless connected, in terms of systems, by energy and information flows that are environmentally sensitive.

Schneider and Sagan proceed to relate the diversity of species to the efficiency of energy use by ecosystems. They posit that

> an increase in species diversity may represent the ecosystem searching out new pathways for energy degradation. As these systems increase their diversity, more pathways are found. This creates redundancy and makes the ecosystems less likely to shut down. By contrast, an

> ecosystem with relatively few species is more vulnerable to mishaps that can compromise or destroy the degrading capacity of the whole system.[103]

Indeed, redundancy in ecosystems follows Luhmann's logic about how autopoietic systems are functionally differentiated, specialized into subsystems. So while Luhmann rather pessimistically describes firstly the difficulty of translating ecological problems into a language that a system can understand, and secondly how to imbue these problems with a larger social meaning, his argument, I suggest, permits us to view ecosystems in the same way as social systems. From this viewpoint, we see that nature's economy is like any other economy, subject to crashing if it becomes too narrow.

Energy flows of ecosystems can also be related to economic flows. Reading cities as dissipative systems, C. Dyke posits that economic trade can be regarded as a flow pattern: 'The import-replacing cities look for all the world like strange attractors – structures that emerge from economic flux, and then begin to organize that flux'.[104] Replaying this idea in the urban centres of Chicago and New York, Dreiser's novel details a thermodynamic economy. From the start, the narrator views the morality of human behaviour as far from equilibrium: 'When a girl leaves her home at eighteen, she does one of two things. Either she falls into saving hands and becomes better, or she rapidly assumes the cosmopolitan standard of virtue and becomes worse. Of an intermediate balance, under the circumstances, there is no possibility'.[105] Described in the novel as 'a giant magnet',[106] Chicago becomes a model of Dyke's 'strange attractor', organizing standards of wealth and virtue. Carrie succeeds because she masters the complex, urbanized trading of virtue for money, learning how to profit from an economic/moral gradient (initially seen in Carrie's willingness to be seduced by the salesman, Drouet, to improve her economic situation). When the setting shifts to New York, this gradient increases rapidly, and so does the complexity of its economic structure, which puts Carrie's lover, Hurstwood, formerly a success in making social connections in Chicago, at a profound disadvantage.

Like all non-equilibrium thermodynamic systems, economic systems produce waste. Dyke comments: 'Trash, soot, and sludge seem an annoying and inconvenient by-product of our lives and activities rather than a necessary feature of them. But without a gradient down which material can cascade, no dissipative structure can remain stable'.[107] In the New York cityscape, where 'spending', for Dreiser, is both economic and metabolic, Hurstwood's downfall illustrates how waste is a human cost paid to maintain the economic flow. In contrast, Carrie's suggested fate (sketched by the image of her rocking in the chair at the end of the novel) is to remain functioning far from equilibrium: she thus embodies the operation of economic systems. For Philipp Schweighauser, she represents the equation between monetary 'investment' and the pull of desire,[108] which in the novel remains unresolved.

In Wharton's novel, the trading of money and morality also operates far from equilibrium. Unlike Carrie, however, Lily does not understand this principle, and her actions try to balance the two, which has predictably negative effects in a social marketplace so interpenetrated by economic systems that '[e]ven fortunes supposed to be independent of the market either betrayed a secret dependence on it, or suffered from a sympathetic affection'.[109] Central to Lily's predicament is her debt owed to Gus Trenor. She tries to repay the debt, but Trenor clearly states that he wants to trade sex for money, a proposal she immediately rejects. As Wai Chee Dimock notes, Lily's attempt at '"[p]ayment in kind", the most primitive form of barter economy, has no place in a highly developed social marketplace, which trades on the putative equivalence between disparate entities'.[110] Like all women in upper-class New York society, Lily remains separate from the male-dominated environment of Wall Street, and so she cannot receive information directly about this environment, restaging Luhmann's system/environment distinction.

At the end of the novel, Lily repays the debt, thus temporarily balancing money and morality, which bankrupts her. That same night, she fatally overdoses on a sleeping agent. In the logic of the social marketplace, repaying the debt removes the gradient, and Lily reaches a deathly equilibrium. It is also a state of maximum entropy, that is indeterminability, for Wharton leaves it unclear whether Lily has taken the overdose deliberately.

Read as thermodynamic narratives, Dreiser and Wharton's novels reveal a somewhat muted but still significant ecocritical perspective. The narrator of *Sister Carrie* comments: 'How dispiriting are the days during which the sun withholds a portion of our allowance of light and warmth. We are more dependent upon these things than is often thought. We are insects produced by heat, and pass without it'.[111] As we follow Carrie's path to fame, we see an increased reliance on artificial lighting (culminating in the dazzling spectacle of Carrie's name in lights) and 'exhibition[s] of showy, wasteful, and unwholesome gastronomy as practiced by wealthy Americans, which is the wonder and astonishment of true culture and dignity the world over'.[112] Likewise, each environment of upper-class wealth that Lily inhabits is more brightly lit than the last, and the conspicuous consumption of resources is nowhere more blatantly apparent than when she accompanies the Gormers, who are 'taking a party out to Alaska in their private car'.[113] The extravagance of their 'nature trip' clearly marks them as ecologically ignorant, their social set reducing the beauty and value of the natural environment of Alaska to a commonplace summer vacation spot, the equivalent of an upper-class resort like Newport in Rhode Island. Lastly, the deaths of both Hurstwood and Lily occur in complete darkness, a reminder of what occurs when ecosystems fail to process sunlight efficiently. These novels help to illuminate the thoughts of Dyke, who believes that

> our existence as dissipative structures defines a space of possibilities for us, and does so rather tightly. We know from the standard thermodynamic analyses of human life that if we conserve and recycle we can lengthen the course that materials follow as they run through our hands. We know that if we use sunlight (and its immediate and inevitable correlates such as wind) we can select a composition of material flow that gives us a longer thermodynamic horizon.[114]

We are hence part of a thermodynamic as well as a monetary/material economy, and this arrangement warrants a careful study of the interconnections between the two that shape our biological and social lives. Yet we should be careful not to re-instill the drive for order that characterizes self-organizing systems, a drive that, as we have seen, closes down the possibilities for a more robust view of ethics.

IV. Social work and moral parasites

What, then, about a drive against (self-)order? Michaels's reading of Wharton's novel, as we have earlier discussed, argues that Lily's failure not only constitutes no resistance to capitalism, but in fact reinforces it: her death makes her even more valuable to social 'collectors' like Selden.[115] Yet interpreting Wharton and Dreiser's logic as a thermodynamic rewriting of the naturalist plot of decline (as this chapter has done) discloses a paradox that hides in plain sight:

for Lily and Hurstwood, things are best *only* before the books actually start. For them, to put it bluntly, the beginning is *already* the end. In Luhmann's view, granted, the making of meaning is itself a paradoxical act. Even so, moments of disruption in both novels are not reconcilable. According to Schweighauser, a scene in Dreiser's novel overrides the ordering of noise (dictated by self-organization) in and by the narrative: Hurstwood's brief employment as strikebreaker, immediately before Carrie leaves him to fend (unsuccessfully) for himself. Schweighauser remarks that 'there is no authoritative narrative voice or character to tell readers how to interpret the strike. It is in the very absence of narrative reintegration that the contradictions of capitalist consumer culture emerge with raw, unmediated force.'[116] From Schweighauser's perspective, we could therefore say that 'literary strategies of representation' that 'manage the threat of social disturbance' through the appropriation of complex, thermodynamic systems behaviour are put into question.[117]

This disruption of order also occurs in Wharton's novel, initiated by Lily's entrapment by a performance that is explicitly *not* her own. Gus pretends to be Judy, his wife, and lures Lily to their house. Here Gus informs Lily that his investments in the stock market on her behalf were in fact a loan, and that she is now in debt to him: a debt which he means her to pay sexually. Although contingent, the system of social regulation, we cannot fail to notice, tightens its grip on Lily. When in desperation Lily visits Gerty, we see how, in the parlance of systems theory, morality becomes free-floating. As William Rasch details, a rather disturbing aspect of modernity, according to Luhmann, is that morality 'has detached itself from its premodern locus in religion and has become a self-replicating, parasitic invader of the various modern, functionally-differentiated social systems'.[118] Consequently morality infects systems codes so that, rather than reflecting the needs and goals of specific systems, codes govern the operation of systems based on the 'binding' determinant of '"good" (or "bad")'. In response, ethics 'emerges as the by-product of a system's attempt to preserve its own reproduction from the ravagings of moral infection'. Thus moral decisions are rewritten as a 'consideration of action in the face of competing alternatives'.[119] Wolfe comments: 'In this way, ethics attempts to ensure the autopoiesis and autonomy of the various function systems by limiting, and, if one likes, relativizing morality'.[120]

Lily's visit to Gerty replays the conflict, earlier addressed, between ethics and morality. In Wharton's novel, morality has indeed become 'parasitic', attaching itself onto another system of social regulation: Gerty's charity, a 'Girls' Club', a locus for class-, race- and gender-constructed ideas of what is socially acceptable. At first, Gerty espouses an ethics of economic responsibility that 'relativizes morality', encouraging Lily's empathetic response to the young women as she makes a donation to her financially 'struggling' organization.[121] Further honing a sense of ethics, Gerty openly espouses a belief that the upper class has a social responsibility in speaking to Selden: 'Do look at Mrs George Dorset's pearls – I suppose the smallest of them would pay the rent of our Girls' Club for a year'.[122] Yet morality intercedes, as Gerty, reflecting the dominant social beliefs of the upper class, disparages Rosedale's rather sizeable contribution.[123] Gerty's class loyalty is of no avail, as the narrator then labels her 'a parasite in the moral order',[124] perhaps to distract us from the growing concern about the change in or confusion over Lily's performance. This tension is signalled by a remark from the narrator (echoing the image of Carrie, in Dreiser's novel, 'surfing' over the poor) about Lily: 'She had always accepted with philosophic calm the fact that such existences as hers were pedestalled on foundations of obscure humanity'.[125] As the truth of the transaction with Gus emerges, Lily's acceptance of upper-class logic negates the ethical issue of Gus's treatment of her and amplifies the moral cost. But such a cost is in turn shifted to the lower-class women of Gerty's charity. Lily asks Gerty, 'There are bad girls in your slums . . . Don't they always go from bad to worse? There's no turning back – your old self rejects you, and shuts you out'.[126] She then says to Gerty: 'I've sunk lower than the lowest, for I've taken what they take, and not paid as they pay'.[127]

In the ambiguous phrasing, the concern is now fully voiced: is Lily a parasite or a prostitute? We could conclude that a parasite would be ethically suspect, while a prostitute would be morally so. In this respect, the matter is settled in Dreiser's novel, as the lack of moral equilibrium is accepted as part of a systems-based ethics – which bears repeating: 'When a girl leaves her home at eighteen, she does one of two things. Either she falls into saving hands and becomes better, or she rapidly assumes the cosmopolitan standard of virtue and becomes worse. Of an intermediate balance, under the circumstances, there is no possibility'.[128] But the matter remains unsettled in Wharton's novel, a deferral routed through the curious (mis)judgement of Gerty as a 'moral parasite'. We might conclude that morality *as* parasite enacts a point of no return for systems, as they become misguided, here shutting down the potential for performativity embodied by Lily in the *tableau vivant* scene. We can thus say that social regulation in Wharton's novel registers as especially destructive when guided by moral concerns (concerns which Bertha and Gus skilfully use against Lily), a situation that attempts to contain, for Wharton, the ghost in the machine – desire – that pressures self-organizing thermodynamic systems.

We must return to the question of parasite or prostitute. Any means of resolving it, in *The House of Mirth*, is guided by the selection for meaning based on ethics or morality. That is, ethically, it would certainly be justified to ask who in Wharton's novel is *not* either a parasite or a prostitute – or both? (This is a question that invariably haunts capitalism.) Morally, the question is easier to answer, but short-circuits the logic of the narrative, so far as we have discussed. To take the moral route, as the novel does, is to introduce a new narrative that parasites meaning. From this perspective, Lily's death exemplifies how self-reference (already parasited by morality, we should note, in that Lily's self-image is shaped by upper-class values) is posed against meaning. Her death scene is staged through an encounter with a young woman, Nettie Struther, whom Lily had previously helped through Gerty's charity. Nettie's outcome, which she relates to Lily, exemplifies the triumph of ethics over morality, for she had a baby out of wedlock then was able to marry a man empathetic to her situation. Lily has throughout the narrative displayed a sense of empathy that has the potential to change her perspective; when she participates in the charity, she becomes 'drawn out of herself'.[129] Thinking about Nettie afterwards, Lily appears to shift from a critical stance of morality to one of ethics: 'It was a meager enough life, on the grim edge of poverty, with scant margin for possibilities of sickness or mischance, but it had the frail audacious permanence of a bird's nest built on the edge of a cliff – a mere wisp of leaves and straw, yet so put together that the lives entrusted to it may hang safely over the abyss'.[130] But in keeping with what I see as the novel's *dazzling* shifts between ethics and morality, Lily circles back to the moral issue in her life: she concludes that she would need the help of an empathetic partner to (re)build her own life. While this other world of ethical possibilities (*alien*, indeed, to upper-class values) seems closed, it opens for the last time in a strange, *otherworldly* dream about Nettie's baby, influenced no doubt by Lily's miscalculation of the dosage of a sleeping agent that proves to be fatal.[131]

Lily's 'big sleep', if you will, leaves ambiguous the connections among and consequences of the systems of social regulation in the novel. Not only does the novel refuse to indicate whether Lily intended to overdose; Selden's account, serving as the 'official' story, challenges by its troubling omissions what the reader has witnessed: 'It was true, then, that she had taken money from Trenor; but true also, as the contents of the little desk declared, that the obligation had been intolerable to her, and that at the first opportunity she had freed herself from it, though the act left her face to face with bare unmitigated poverty'.[132] What this upper-class-endorsed account does is shift the blame, as Ellen J. Goldner explains, from the interdependence of economic systems responsible for the unregulated speculation that creates financial crisis to the moral conduct of unmarried women.[133] Yet the final meaning of Wharton's narrative is indeterminate; it is, again in the parlance of systems theory, contingent upon the reader's own reference. It is once more a doubling back, a doubling up of the meaning system, here forced, we might say, into oscillating between perception and communication, mimicking how social media pretend to offer second-order observations in the guise of first-order observations (a theme that also appears in the ending of *The Custom of the Country*).[134]

If such moments of narrative disruption offer *an-other* cultural perspective that is or marks the 'outside' of the system, we might investigate how this shift recharges the social environment. The threat of social violence in *Sister Carrie* does not go away completely after the strike scene, for it returns in the hint of social agitation in the scene where Hurstwood, all the more desperate and socially marginalized, takes part in the 'army of tramps' march. Organized by 'an ex-soldier turned religionist',[135] the march stages a collective spectacle of a demand for charity. Similarly, Lily's thoughts and surreal dream about Nettie invoke a 'fragile' safety set of social help networks that offers the start of a systematic approach to help, or, as Gavin Jones puts it, 'public moves that institutionalized relief',[136] for the lower classes. These images of ideological 'social work', in both novels, frame Jones's account of how social factors, such as immigration, catalysed 'the development of voluntary social work – inspired by powerful beliefs in the "social gospel of reform" – and of professional sociology, which strove to define and quantify the extent of poverty'.[137] In his reading, Jones traces a discourse of 'pauperism', which states that some are afflicted with or by a state of helplessness that cannot be overcome or reformed. Hurstwood's addiction to comfort in the later chapters of the novel is easy to see; yet Jones claims that Lily's addiction to the sleeping agent is also a sign and symptom of pauperism. Hurstwood's relatively painless death (suicide by gas), Jones asserts, thus represents his own 'solution' to the problem of pauperism.[138]

What we therefore see in the two novels is a vision of a second form of Goldner's idea of 'interdependence' (one granted even less social visibility), that is, system(at)ic, class-based poverty cast as individual failure in the bodies of Lily and Hurstwood, accentuated by a moral/tragic dimension of human behaviour: in sum, the discourse of pauperism. As Jones contends, 'Wharton conceives of poverty as a broad and complex state of marginalization . . . that allows for an alignment of gender and class much more radical than most critics would allow'.[139] Perhaps it could be said that, while waste may figure in thermodynamic terms as part of a system's autopoiesis, in human form it also creates the impetus for a state system of social care. What, over time, comes into view is a nascent form of biopower, described by Livingston as 'political control over the fundamentals of life',[140] an idea which will be more extensively explored in the form of water management in Wharton's *The Custom of the Country*.

While, to a degree, the two novels provide another way of looking at globalization, as the normalization of class-based discrimination or exploitation across race and gender tropes, they do question the connection of state power (domestic reform programmes) to the emerging imperial global presence discussed in the previous chapter. This is because both novels, in their plots of decline, perforce depict dissipative systems. Thus, according to Livingston, not only can biopower trace and then contest the hegemonic worldview that informs how state control over life is rationalized; dissipative systems also move us away from 'the notion of systems that "perpetuate themselves" and [patrol] their boundaries', the analogue for 'what patriarchal nations and empires, and the subjectivities and systems modelled in their image, seek to do'.[141] What we get in the last image of Hurstwood's self-annihilation, his permanent dissipation, is the vision of a programme not state-determined per se, yet still chilling. As Jones, aware of the mounting ironies, remarks: 'Nobody would be turning on the gas in an official capacity; this would not be genocide, technically speaking. More conveniently still, in this world of "moderate" organization the conditions of poverty would be allowed to take care of themselves'.[142] Just as Lily's death clouds cause and effect, so does this (deathly) fantasy, a contingency that emerges from the same paradoxical logic that constitutes dissipative systems.

3

THE ECOLOGY OF EMPIRE

In the early twentieth century, Edward L. Thorndike wrote a number of articles for *Popular Science Monthly* in which he critiqued ideas of animal intelligence, provoking responses from his readers. These responses consisted of observations of animal behaviour (for example, a reader would describe a response from his/her dog that suggested signs of intelligence), which Thorndike would then challenge by citing evidence from experiments.

FIGURE 3 'Jack and his master', from the article 'Do animals reason?', *Popular Science Monthly*, 5/46 (1899)

In 'The evolution of the human intellect', published in November 1901, he writes his most comprehensive statement of his position in the debate about animal intelligence. He argues for a biological continuity between the mental processing of humans and animals, in contrast to William James (cited in the article) who posited that there was a 'break' that could not be surmounted. At the same time, Thorndike once more asserts human superiority: 'There may be a few scattered ideas possessed by the higher animals, but the common form of intelligence with them, their habitual method of learning, is not by the acquisition of ideas, but the selection of impulses'.[1] That is, animals use a trial and error approach, what Thorndike calls a process of 'selection', whereby '[t]he profitless acts are stamped out; the impulses to perform them in that situation are weakened by reason of the positive discomfort or the absence of pleasure resulting from them'.[2] While humans also often use such an approach, signifying an evolutionary connection with animals, their 'associations . . . between situation and act' achieve an optimum level. As a result, humans appear more advanced, but not fundamentally separate from animals: 'Indeed it may be that this very reason, self-consciousness and self-control which seem to sever human intellect so sharply from that of all other animals are really but secondary results of the tremendous increase in the number, delicacy and complexity of associations which the human animal can form'.[3]

When Thorndike does suggest that humans have undergone a profound evolutionary advancement, he couches his suggestion in the language of continuity: 'an apparently new type of mind results, which conceals the real continuity of the process. This mental evolution parallels the evolution of the cell structures of the brain from fewer and simpler and grosser to many and complex and delicate'.[4] Like Spencer, Thorndike tends to rely on a progressive model of evolution that acts in uniform ways across species. Finishing his argument, Thorndike revisits the similarities between humans and animals, before arriving at a conclusion that supports human superiority:

> Nowhere more truly than in his mental capacities is man a part of nature. His instincts, that is his inborn tendencies to feel and act in certain ways, show throughout marks of kinship with the lower animals, especially with our nearest relatives physically, the monkeys. His sense powers show no new creation. His intellect we have seen to be a simple though extended variation from the general animal sort. This again is presaged by the similar variation in the case of the monkeys.

> Amongst the minds of animals that of man leads, not as a demigod from another planet, but as a king from the same race.[5]

Here Thorndike restages the pronouncement of Darwin that the crucial differences between humans and animals are of degree and not of kind.

In *The Call of the Wild* (1903) and *White Fang* (1906), Jack London supports the idea of biological kinship and limits on animal intelligence. He denies his canine actors the power of abstract thinking and evidence of complex reasoning, even using some of the same language as Thorndike, such as repeated reference to impulsive reactions. Yet London remains suspicious of claims about human superiority, especially when voiced through the notion of kinship. His dog novels dramatize how the struggle for survival, an evolutionary precept, connects humans and dogs, each having qualities that the other species needs. In his depiction of cooperation rather than competition between men and dogs in the harsh Yukon environment, London challenges the rationale for the exploitation of animals, the idea of human dominance based on the ability to think in complex and abstract ways.

London's depiction of survival, however, reflects a masculine drive for power and mastery that fuels both imperial desires for national expansion and the global exploitation of natural resources. Despite his fluid model of evolution that destabilizes identity, the logic that upholds human superiority returns. It is a return that London cannot see, much less address, due to the narrow ethical parameters of his dog narratives. Two letters from London, in June 1905, to George P. Brett, his long-time editor at Macmillan, encapsulate this return. In a letter of 7 June, London exults over his purchase of '130 acres of the most beautiful, primitive land to be found anywhere in California'.[6] Then in a letter thirteen days later, he remarks: 'It was one of those bargains that one stumbles on to in much the same way that one finds a gold mine'.[7] The description of his land purchase, occurring in between the publication of *The Call of the Wild* and *White Fang*, discloses a tendency in the two novels to channel ecological concerns through a liberal humanist ideology that justifies human dominance over the natural landscape. Despite his depiction of gold-mining as environmentally destructive in 'All Gold Canyon', written in the same year as his letters to Brett, London remains committed to a narrative of animal domestication that echoes the taming of the wild frontier, a process that transforms nature into a resource.[8] At the same time, however, a counter-narrative of biological kinship emerges in the two novels that depicts an ongoing co-evolution of humans and their companion species, dogs. What interests me is how the conflicts between these narratives question the boundaries between nature and culture and destabilize a national subjectivity. While always a fragile concatenation of gender, race and class signifiers, the national subject in the early twentieth century was fragmented both by a global campaign of empire building, fuelled by fantasies of frontier conquest, and by the closing of the physical frontier.

The Call of the Wild and *White Fang* are particularly instructive, I believe, for observing the complex linkages among evolution, citizenship and subject making. Driven by imperialism, national progress implied that Native Americans were doomed to lose out in an evolutionary struggle. But such a loss reopened questions about their apparent location, in nature and/or culture, due to their unresolved status as citizens. The failure to deal effectively with the issue of Native American citizenship threatened to reveal something inauthentic at the base of the US subject. London's dogs, as we will see, re-enact anxieties about race and citizenship through a masculine iconography of performance that enacts ghostly transformations in the imagined spaces of the Northland (the Yukon territories) and Southland (California). Thus the novels suggest that the restoration of the national subject is a fantasy whose arrival is always in question.

Using systems theory to explore the biological politics of the frontier, this chapter examines how dogs as actors in history negotiate between liberal humanist ideals of self-possession and evolutionary loops of self-making. In the Northland and Southland, where gender, race and class are always in play and open to revision through cross-species relationships, ecological issues are expressed through national subject making. The narrative trajectory of *The Call of the Wild* and *White Fang* posits that there is no escape from troubling questions about national expansion, as acts of imperial violence produce a (self-)compelled performance of citizenship. I argue that what Judith Butler calls the 'horizon of social being' is intimately connected, through London's vision of biological kinship, to global spheres of influence.[9]

I. *The Call of the Wild* and the national frontier

Frontiers have long been sites for the production of national subjectivity. But by 1890, the frontier, as Richard Slotkin demonstrates, 'became primarily a term of ideological rather than geographical reference'.[10] As historian Frederick Jackson Turner points out, that same year marked the closing of the frontier. While there were still 'various "resource frontiers," in the timber ranges, the new oil fields of Texas and Oklahoma, and the Alaskan gold rush', Slotkin explains that 'individual entrepreneurs and settlers on the new frontiers of the industrial and railroad eras had to contend (on increasingly disadvantageous terms) with large capitalist enterprises'.[11] The traditional definition of the frontier as a site for the making of rugged male individualism had to be rethought in order to reconcile masculine heroic feats with the economic dictates of capitalism. How then could the US progress economically without losing the sense of the wild that defined the national male subject?

Theodore Roosevelt's 1899 speech, 'The Strenuous Life', emphasizes the need to find an answer to this question. He warned that 'the nation that has trained itself to a career of unwarlike and isolated ease is bound, in the end, to go down before the other nations which have not lost the manly and adventurous qualities'.[12] Jennifer L. Fleissner posits that the revival of masculine outdoor adventure becomes the perceived polar opposite of feminine domesticity, and remarks: 'Given the locales it tends to favor (not to mention its frequent link to imperialist ambitions), we might consider calling this remasculinizing therapy "the west cure" '.[13] In *The Call of the Wild*, the opening scene in the civilized Southland tends to support this belief. At the luxurious ranch of Judge Miller, where the canine protagonist, Buck, resides, the natural landscape has been mapped and tamed in the form of 'graveled driveways which wound about through wide-spreading lawns' and 'an endless and orderly array of outhouses, long grape arbors, green pastures, orchards, and berry patches'.[14] Although Buck stands out from the house-dogs, 'Toots, the Japanese pug, or Ysabel, the Mexican hairless – strange creatures that rarely put nose out of doors or set foot to ground', he evokes a lost male vitality: 'He had lived the life of a sated aristocrat . . . [he] was ever a trifle egotistical, as country gentlemen sometimes become because of their insular situation'.[15] The image of Buck as 'sated aristocrat' reflects anxieties that the economic prosperity brought about by corporate interests will erode a sense of identity as a democratic nation and impede national growth. As Slotkin elaborates, Turner's 'nostalgia' for the lost frontier 'implies a critique of the corruptions of the present. The Frontier of the past appears as the place in which, once upon a time, the political and social life of a European people was transformed, morally regenerated, and given a distinctively democratic direction'.[16] Represented by Buck's regal airs, the Southland environment has undone the frontier transformation that has built the national character.

But Buck's 'insular situation' is far from permanent, or absolute. A gardener's helper, Manuel, steals Buck and sells him to people who need working dogs in the Klondike. Thus starts Buck's journey into wild nature, which London associates with Buck's constitution, for he 'saved himself by not becoming a mere pampered house-dog. Hunting and kindred outdoor delights had kept down his fat and hardened his muscles'.[17] As a working dog, his wildness is an asset. Referring to 'the mythology of Buck', Vicki Hearne thinks that a 'dog or horse trainer's special interest in "wild and uncontrollable" animals assumes that the degree of violence in a dog's resistance to authority may indicate the depth and power of response such a dog might give when the gap between command and obedience is closed'.[18] Supporting Hearne's belief about the potential for 'resistance to authority', the narrative trajectory in *The Call of the Wild* is informed by the theory of atavism, an evolutionary regression to a former hereditary condition of wildness. The novel's opening epigraph, taken from the first quatrain of John Myers O'Hara's poem 'Atavism', portrays evolution as a naturally regressive force: 'Old longings nomadic leap,/Chafing at custom's chain;/Again from its brumal sleep/Wakens the ferine strain.'[19] In contrast to Herbert Spencer, who believed that the survival of the fittest meant that progress was naturally assured, London takes the perspective of T. H. Huxley, who questions this belief:

> I suspect that this fallacy has arisen out of the unfortunate ambiguity of the phrase 'survival of the fittest'. 'Fittest' has a connotation of 'best', and about 'best' there hangs a moral flavour. In cosmic nature, however, what is 'fittest' depends upon the conditions . . . [I]f our hemisphere were to cool again, the survival of the fittest might bring about, in the vegetable kingdom, a population of more and more stunted and humbler and humbler organisms, until the 'fittest' that survived might be nothing but lichens, diatoms, and such microscopic organisms.[20]

London's ideological frontier, the Northland, thus becomes an experimental space, an outdoor laboratory to test the effects of atavism on Buck.[21] According to Lawrence J. Berkove, London was aware of the 'dispute' between Huxley and Spencer; he 'admired it for the vigor and life of its intellectual battle'.[22] In the experimental space of the

Northland, London therefore theorizes about how changing conditions might redefine biological fitness and how Huxley's predictions might be validated through a domesticated dog's struggle for survival, a struggle undistorted by moral concerns. This speculation about evolutionary theory is informed by London's scientific fascination with violence. In a letter of 23 February 1902 to his friend Cloudesley Johns, London claims that the execution by hanging of a man 'was one of the most scientific things I have ever seen'. He describes the event from a detached, clinical point of view, noting that it took 'fourteen and one-half minutes for the heart to run down . . . ⅕ of a second elapsed between the springing of the trap and the breaking of his neck & severing of his spinal cord'.[23]

Scientific violence emerges in *The Call of the Wild* in scenes of brutality to animal bodies that figure in what Seltzer terms 'struggles to make interior states visible: to gain knowledge and mastery over bodies and interiors by tearing them open to view'.[24] At the start of Buck's training, the man in the red sweater physically 'breaks' him by beating him with a club until Buck is too exhausted to retaliate. Thus the lesson of discipline is literally beaten into his body and the bodies of other dogs: 'He saw, once for all, that he stood no chance against a man with a club . . . Again and again, as he looked at each brutal performance, the lesson was driven home to Buck: a man with a club was a lawgiver, a master to be obeyed, though not necessarily conciliated'. Buck's intelligence is one component of his fitness for the Northland, as he realizes he must respect authority without however submitting blindly to it, and that rebellion can mean death: 'He was beaten (he knew that); but he was not broken . . . [H]e did see beaten dogs that fawned upon the man, and wagged their tails, and licked his hand. Also he saw one dog, that would neither conciliate nor obey, finally killed in the struggle for mastery'.[25]

If London regards evolution as a natural force that determines subjectivity, then dogs, seen as bodies of nature, also help to sustain gender definitions. This process fits with Cary Wolfe's discussion of 'rewriting of the problematic of gender in terms of species'.[26] That is, for London, questions about how behaviour expresses and sustains gender are rewritten at the level of species, making gendered acts appear more instinctive. Scientific violence shapes how Buck's training disciplines his body to fit the masculine economy of the Northland and how the sacrifice of animal bodies, as an expression of cultural values, underwrites this economy.

That sacrifice is central to a masculine economy is illustrated through the next lesson that Buck learns upon arriving in the Northland. Through his journey that London describes as 'from the heart of civilization . . . into the heart of things primordial,' Buck observes the savage way that husky dogs, described as 'wolfish creatures', fight through the death of Curly, a female Newfoundland. She makes a friendly overture to a husky that is 'the size of a full grown wolf'. Without warning, the husky's attack on her demonstrates 'the wolf manner of fighting, to strike and leap away'.[27] She tries to fight back, but is soon knocked off her feet and torn apart by the dogs that have gathered to watch the battle. London describes Buck's grasping of this lesson: 'The scene often came back to trouble him in his sleep. So that was the way. No fair play. Once down, that was the end of you. Well, he would see to it that he never went down'.[28] Through masculine fantasies of violence, this scene, of which Buck repeatedly dreams, connects survival and identity. To survive, that is, means not acting like Curly. Which, in turn, means that females are at the bottom of the gender hierarchy that structures biological survival. In a letter of 23 June 1899 to Johns, London insists:

> Remember, there is even a higher logic than moral or formal logic. Moral and formal logic demonstrate thoroughly that woman should vote; but the higher logic says she shall not. Why? Because she is woman; because she carries within her that which will prevent, that which will no more permit her economic and suffragal independence, than it will permit her to refrain from sacrificing herself to the uttermost to man.[29]

Not only does London elevate cultural belief to natural status; he misreads what the survival of the species implies and avoids asking why females would be less inclined to prioritize their own survival and more willing to see their ultimate purpose as being victims.[30] What Buck learns, then, is the lack of value females have in a masculine economy, a lesson that authorizes male dominance.

These two lessons, the scene with the man in the red sweater and Curly's death, illustrate 'the law of club and fang', London's reinvention of the survival of the fittest doctrine that strips away all civilized traits, such as fairness, compassion and honesty. In a systematic devaluing of these traits, Buck, in order to survive, learns to steal food and fight with no mercy, as the wolves do. Again echoing Huxley, London suggests that Buck's survival in the rugged Northland environment resists the idea of an evolutionary progress: 'His development (or retrogression) was rapid'. The disciplining of Buck is again depicted through the body:

> He achieved an internal as well as external economy. He could eat anything, no matter how loathsome or indigestible; and, once eaten, the juices of his stomach extracted the last least particle of nutriment; and his blood carried it to the farthest reaches of his body, building it into the toughest and stoutest of tissues.[31]

London emphasizes that his 'development' is not thought out, for 'instincts long dead became alive again'. Buck's development traces hereditary patterns of violence:

> The domesticated generations fell from him. In vague ways he remembered back to the youth of the breed, to the time the wild dogs ranged in packs through the primeval forest and killed their meat as they ran it down. It was no task to learn to fight with cut and slash and the quick wolf snap. In the manner had fought forgotten ancestors. They quickened the old life within him, and the old tricks which they had stamped into the heredity of the breed were his tricks. They came to him without effort or discovery, as though they had been his always.[32]

The attempt to dramatize evolutionary regression ends with the remark: 'Thus as token of what a puppet thing life is, the ancient song surged through him and he came into his own again; and he came because men had found a yellow metal in the North, and because Manuel was a gardener's helper whose wages did not lap over the needs of his wife and divers copies of himself'.[33] Evolution is depicted as a natural force that both guides individual development and acts on animal and human populations. Yet in London's account, masculinity is both natural and man-made, allying a naturalistic determinism with an evolutionary atavism and economic progress, symbolized by the gold rush.

To some degree then, masculine values are overdetermined, the narrative trajectory staging through gendered bodies a struggle both for leadership and survival. Buck's first clash with Spitz, the leader of the dog team, is interrupted by a surprise attack from a pack of 'starving huskies . . . who scented the camp from some Indian village', and it is this event that 'projected their struggle for supremacy far into the future'.[34] Afterwards the team is so visibly in disarray that the courier fears the dogs might go mad. While the starving dog pack, associated with a Native American otherness, interferes with Buck and Spitz's battle for dominance, the death of Dolly revives it. She goes mad, confirming the courier's concerns about the team's collective lack of mental fitness, and suggesting that as female she appears weaker and more likely to falter. Also implied is that females threaten masculinity, as Dolly turns on Buck. Her madness, furthermore, repeats the undisciplined wildness of the huskies that embodies the racial trait of Native American 'savagery'. White male discipline must be preserved in order to contain the 'race problem' that London references in his letter of 23 June 1899: 'The different families of man must yield to law – to LAW, inexorable, blind, unreasoning law, which has no knowledge of good or ill, right or wrong'.[35] Accordingly, Buck is saved by the dog-driver who kills Dolly. Spitz takes advantage of Buck's weakened condition and attacks him, only to be repelled by the dog-driver. Acting as commentators, the two men in charge of the dog team speculate on who will win the final showdown. Again the man in the red sweater's lesson is invoked, which allows Buck to equal Spitz in 'strength, savagery, and cunning' through the sacrifice of the female body of Dolly.[36] The repetition of the lesson, central to the making of Buck's subjectivity, signifies the multiplicity of his performance. According to Butler: '[P]erformativity is thus not a singular "act", for it is always a reiteration of a norm or set of norms, and to the extent that it acquires an act-like status in the present, it conceals or dissimulates the conventions of which it is a repetition'.[37] Repeating the codes of white masculinity, Buck's performance acts out evolutionary struggle and selfhood.

Buck's training shows how masculine subjectivity and ambition are reinforced through bodily violence. Although Jonathan Auerbach perceptively questions London's logic of work, the scenes of Buck as part of a team of dogs pulling a mail-sled appear to matter less in terms of valuing work and more in reinforcing the themes of individualism, atavism and masculine ambition.[38] When Buck fights with Spitz, his rival for leadership of the team, the narrator remarks that Buck does so 'because it was his nature, because he had been gripped tight by that nameless, incomprehensible pride of the trail and trace'.[39] While Buck's atavism allows him to match the savagery of Spitz, his imagination, shaped by evolutionary processes, enables him to win and thus demonstrate his masculine will to power. Buck tricks Spitz into believing that he is rushing for his shoulder and instead attacks his leg and breaks it with his teeth. After Spitz is tricked again, he cannot stay upright on two broken legs. Buck knocks him over and kills him.

Even with the erasure of female animal bodies, London still finds a way to introduce the supposed dangers of domesticity.[40] Described as 'lacking the wonderful patience of the trail', the three Southland travellers who purchase the dog team embody the complete opposite of Buck's disciplined male economy.[41] To further demonstrate this difference, we first see the travellers from Buck's point of view: 'When driven with his mates to the new owners' camp, Buck saw a slipshod and slovenly affair, tent half stretched, dishes unwashed, everything in disorder; also he saw a woman'.[42] The woman is little more than a female stereotype of fussy behaviour, self-indulgence and incomprehension of outdoor living. First, London suggests that the two men, Charles and Hal, are 'manifestly out of place', then he indicates that overly civilized attitudes will obstruct the daily routines in the Northland: 'That Hal's views on art, or the sort of society plays his mother's brother wrote, should have anything to do with the chopping of a few sticks of firewood, passes comprehension'.[43] Auerbach states:

> Confronted with an alien environment, the overcivilized family registers chaos, whereas the state of wildness clearly depends on strict regimentation, again only possible through regulated work. Given the absence of such service, at once ennobling and enabling, nature can only be represented by what it is not.[44]

But the forces of nature soon take revenge, telegraphed by London's image of a living earth responding to seasonal change: 'The ghostly winter silence had given way to the great spring murmur of awakening life. This murmur arose from all the land fraught with the joy of living.' As nature cycles through change, frozen rivers begin to thaw. Forcing the dog team to continue across dangerous ice, the hapless travellers arrive at the 'mouth of White River', where an experienced outdoorsman, John Thornton, has made his camp.[45] After they ignore Thornton's warnings about the thin ice, Buck, whose evolutionary reversion has sharpened his instincts, refuses to go any further. Thornton comes to his rescue by cutting him free. Soon thereafter, the travellers and rest of the dog team plunge through the ice to their deaths. Hence the problems caused by a supposedly feminized civilization vanish in a natural disaster that restores the values of white masculinity.

Thornton's intervention allows London to test the effects of love on Buck. His wild nature connected to his 'unbreakable' heart, Buck previously had 'a working partnership' with the Judge's sons, and with the Judge 'a stately and dignified friendship'.[46] Through the relationship of Thornton and Buck, London considers how the law of club and fang might be modulated. London sets up this modulation through the bodies of Thornton's dogs, Skeet and Nig, which echo the gender-race patterns that inform Buck's winning the leadership of the dog team. Seeming to have naturally 'the doctor trait which some dogs possess', Skeet tends to Buck's wounds, the narrator explaining that he is too weak to 'resent her first advances'. Nig, 'a huge black dog', his name an obvious (and of course derogatory) racial reference, with Skeet, gets Buck involved in 'all sorts of ridiculous play' with Thornton, whereby play becomes love play.[47] Buck is compelled to love Thornton because he is an 'ideal master', for he understands the masculine code of 'rough love', calling Buck 'ill names that to Buck were love names'. In his return of affection, Buck conflates love and pain in his 'feigned bite'. Yet London makes it clear that Buck has 'retained his wildness'; while he refuses to steal from Thornton, he will steal from anyone else. Unlike Skeet or Nig, he will not shamelessly seek out Thornton's affection, reminding us of Buck's higher position in the evolutionary hierarchy.[48]

And Buck's most prestigious feat, winning a bet made by Thornton by pulling a thousand-pound sled-load, is ostensibly accomplished out of love: Buck 'felt in some way that he must do a great thing' for his master. Before the contest begins, London devotes an almost voyeuristic paragraph to describing Buck's masculine economy: 'He was in perfect condition, without an ounce of superfluous flesh, and the one hundred and fifty pounds that he weighed were so many pounds of grit and virility'.[49] London's description resonates with the same energies surrounding the turn-of-the-century male icon, Eugen Sandow, the 'Perfect Man', who restored a masculinity threatened by industrialization and modernization, achieving fame through a self-disciplined and scientific regimen of bodybuilding. John F. Kasson writes of Sandow: 'His body's combination of well-proportioned, articulated, and extraordinary measurements fascinated trained observers and novices alike'.[50]

As with Sandow, the spectators are thrilled by Buck's body, reaching out and touching it; both Sandow and Buck distinguish themselves

FIGURE 4 Eugen Sandow: photograph by George Steckel

by meeting physical challenges.[51] That Buck performs before 'several hundred men' emphasizes Kasson's point that, at the turn of the twentieth century, 'the categories of heterosexuality and homosexuality did not squeeze so tightly as to inhibit a man's frank admiration of another man's body'.[52] After Buck wins the bet, the excitement over his bodily feat provides an occasion for male bonding: 'Men were shaking hands, it did not matter with whom, and bubbling over in a general incoherent babel'.[53]

Conversely, in *White Fang*, as we will now observe, evolution runs instead in a progressive direction, transforming the dictates of perfect manhood into the socialization of white bodies. As a critical part of this process, love between companion species (human and dog) is depicted as both a natural and unnatural act. Consequently, the regulation of desire fits within a masculine 'law' that underwrites the code of self-discipline.

II. *Wild Fang* and the ideology of domestication

White Fang opens with London's imaginative rephrasing of Roosevelt's belief that masculine revitalization depends on confronting the wilderness:

> [L]ife is movement; and the Wild aims to destroy movement. It freezes the water to prevent it running to the sea; it drives the sap out of the trees till they are frozen to their mighty hearts; and most ferociously and terribly of all does the Wild harry and crush into submission man – man who is the most restless of life, ever in revolt against the dictum that all movement must in the end come to the cessation of movement.[54]

In his interpretation, Seltzer comments that 'the mastery of men – the *un*natural mastery that revolts against the natural law of "the Wild" that freezes motion – consists of technologies of flow and continuous motion'.[55] Seltzer's argument about how London's writing embraces technological modernization relates to the taming of White Fang, the wolf protagonist, that features his exploits in the modern city of San Francisco. For London, the issue is how to make masculinity appear civilized, while retaining its natural vitality. Accordingly, he follows Roosevelt. As Slotkin explains, rather than offering a frontier thesis as social critique as Turner does, Roosevelt believes that 'the present social order, despite its evident dangers and corruptions, represents the culmination of civilized progress'.[56] Thus London depicts White Fang's development as an evolutionary progression shaped through the civilizing force of love. This strategy unfolds through the joining of episodes of struggle for survival, featuring both human and animal, to scenes that illustrate, through the invocation of a civilized masculine economy, 'the business of love'.[57] Interconnected images of love as informed by violence are suggested by 'the love-making of the Wild, the sex-tragedy of the natural world' and 'the love-tale red-written on the snow'.[58] The scenes of fighting, where the wolf style of attacking without warning contrasts with that of the dog, display masculine attitudes through 'the enmity that the domestic dog feels for the wolf'.[59] That is, 'domestic' is connected to the not- or less-masculine, as White Fang, the wolf, symbolizes a more masculine freedom in a wild, outdoor space. The hierarchy that London maps out is guided by masculine dominance. White Fang follows the rule

> to obey the strong and oppress the weak. Gray Beaver was a god, and strong. Therefore White Fang obeyed him. But the dog younger or smaller than himself was weak – a thing to be destroyed. His development was in the direction of power.[60]

London does allow for an acceptable form of domestication in the guise of disciplined values; like Buck, White Fang pulls a dog-sled for his first master, the Native American, Gray Beaver, where

> he worked hard, learned discipline, and was obedient. Faithfulness and willingness characterized his toil. These are essential traits of the wolf and the wild-dog when they have become domesticated, and these traits White Fang possessed in unusual measure.[61]

The connection of a masculine work ethic with natural traits parallels rules about ownership: 'the law of property and the duty of the defence of property'.[62] That White Fang must learn this law as part of his evolutionary progress naturalizes the cultural values of a patriarchal society. In a parallel with the perfect manhood of Sandow, his whiteness, a 'sign of racial purity', is reinforced through his cultural assimilation, a disciplining of the body through work.[63] He already seems ready for entrance into a socialized white male regime (which we will see in a moment), his 'unusual' reserves of 'faithfulness and willingness' pointing the way.

White Fang's owner after Gray Beaver, the ironically named Beauty Smith, is an example of nature justifying the supposedly negative qualities of unmanliness. That is, London writes, 'Nature had been niggardly with him. He was a small man to begin with; and upon his meagre frame was deposited an even more strikingly meagre head.' In the Northland, he retains domestic duties, 'the cooking for the other men in the fort, the dish-washing and the drudgery'.[64] Of course, White Fang is suspicious of him. London makes White Fang's dislike a natural instinct:

> The good stands for all things that bring easement and satisfaction and surcease from pain. Therefore the good is liked. The bad stands

> for all things that are fraught with discomfort, menace, and hurt, and is hated accordingly. White Fang's feel of Beauty Smith was bad.[65]

Smith's brutal treatment of White Fang is depicted as against the rules of proper male discipline, culminating in a scene in which White Fang fights a bulldog that appears to parody the male-bonding bet in *The Call of the Wild*. Rather than an expression of affection between owner and dog, Smith mocks White Fang when he starts losing, then kicks him. And instead of an expression of community, no one offers to help Weedon Scott when he rescues White Fang from the bulldog's jaws. There is only masculine violence, with Scott punching Smith and threats exchanged among the bulldog's owner, Scott, and Smith.

In taking ownership, Scott must control White Fang's wildness. While White Fang's attack on one of Scott's dogs is tolerated, his attack on Matt, Scott's dog-musher, prompts Scott to say: 'It's all right about the dogs, but we must draw the line somewhere'.[66] His comment highlights the conflict that London attempts to mediate, that is, how masculinity, heretofore connected with wildness, can be preserved through the taming of wildness. Although White Fang can dominate his fellow species members, he must have self-discipline, realizing his place within the white male social order. This idea is once more anticipated by London's 23 June 1899 letter: 'If no race has the moral right to survive in the destruction of lesser races, where, in the whole animal kingdom are you going to draw the line?'[67] The answer that London proposes, as we have seen earlier in the letter, is a law that appears before and beyond culture. While White Fang's wildness threatens London's law of white masculinity, his intelligence makes him a 'higher' animal on the evolutionary hierarchy, a standing marked by *his* whiteness that signifies an ability to adapt to the dictates of civilized behaviour. Like Buck, he stands apart and thus has value within a masculine economy; as Matt says, 'that dog's too intelligent to kill'.[68] Then London suggests that for White Fang, 'the lordship of man was a need of his nature'.[69] For a better, nobler man than either Gray Beaver or Beauty Smith, White Fang must adapt to a new living arrangement, one governed by love rather than hate. As a result, his wild traits are modulated. For example, despite being tamed, he 'never outgrew' his growl, but now it contains a 'new note', one that only Scott's 'ear and sympathy were fine enough to catch'.[70] Accordingly, London depicts White Fang's growing affection for Scott as the 'evolution of *like* into *love*'.[71] The narrative intends to depict love as a natural trait; Darwin proposes that the 'lower animals' exhibit the 'same principle of pleasure' as humans.[72] But at the same time it is an unnatural trait, for Seltzer contends: 'Learning to love pain and the god-like hand of his master, White Fang learns to love at once the pleasure of unnatural acts (acts contrary to every "mandate of his instinct") and the pain of turning from "the natural" to "the cultural"'.[73] London coins a term to depict the evolving of affection from a canine perspective; throughout, Scott is known as the 'love-master'. The centrality of the scenes between White Fang and Scott leads Marjorie Garber to insist that in the novel, 'love . . . is something a dog feels for a man'.[74] In the relationship, White Fang's submissive role, traditionally associated with femininity, shows how, according to Eve Kosofsky Sedgwick, love between White Fang and Scott, both male characters, is expressed through an early-twentieth-century definition of desire 'as the current that runs between one male and one female self, in whatever sex of bodies these selves may be manifested'.[75] Thus White Fang acting as female contains the queer possibilities of love within conventional gender arrangements.

A further conflict, however, arises when Scott must decide whether to take White Fang to California with him. Scott expresses concern about the legal codes that constitute civilized society and the possible consequences of their violation: 'If he didn't bankrupt me with damage suits, the authorities would take him away from me and electrocute him'.[76] After White Fang somehow realizes that he might be left behind, Matt remarks: 'From the way he cut up the other time you went away, I wouldn't wonder this time but what he died.' Scott then tells Matt to 'shut up', exclaiming: 'You nag worse than a woman'.[77] Although London attempts with his parody of stereotypical female behaviour to deny that the situation is at all melodramatic, in reality, it *is* a masculine drama about wildness and civilization. The tamed wolf may die by learning to accept and depend on the codes of male comradeship, the same codes that Scott will violate if he leaves White Fang behind. On the other hand, there is the possibility that civilized society cannot accept the masculine wildness that White Fang embodies.

Scott finally decides to bring White Fang, and they stop in San Francisco en route to their final destination, Sierra Vista. For the dog, 'never had the white men seemed such marvelous gods'.[78] In the novel, San Francisco serves as the Western version of the 'White City' at the 1893 Chicago Exposition, 'representing the pinnacle of Euro-American civilization' as Slotkin puts it.[79] Just 'six miles from downtown Chicago', where Sandow would make a memorable staging of perfect manhood at the Trocadero Theatre, the Exposition also featured an 'Anthropology Building', which, as Kasson notes, displayed 'types from prehistoric times to the present, beginning with North American Native peoples'.[80] London's novel replicates the notion of racial and cultural progress in White Fang leaving Gray Beaver and the Northland behind for San Francisco. The city allows London to subordinate wild to civilized, while the narrator declares: 'All this was the manifestation of power. Through it all, behind it all, was man, governing and controlling, expressing himself, as of old, by his mastery over matter. It was colossal, stunning'.[81] The hyperbolic tone of this passage connects civilization with masculine force; the taming of the wild associated with White Fang becomes the purest expression of masculinity. The demands of the man-made, urban environment, represented by the city looming above White Fang, render him almost completely submissive to Scott: 'As never before, he felt his dependence on the love-master, close at whose heels he followed, no matter what happened never losing sight of him'.[82] In the California scenes, London relates the taming of White Fang to the natural process of environmental adaptation. As White Fang acclimates to Scott's home at Sierra Vista, he learns the laws of patriarchal society, a refined version of the rules of ownership learned while Gray Beaver was his master: '[H]e never regarded the members of the family in any other light than possessions of the love-master'.[83]

Yet there are opportunities to pressure the crucial links between ownership of property and self-possession. In the images from *The Call of the Wild* and *White Fang* that I will discuss, evolution turns subject making into a loop of self-making.

III. The multiplicity of animal bodies

Whether in the wild Northland or civilized Southland, evolution remains unstable, as suggested by London's remark that 'life is movement'. Such an idea anticipates the dynamic view of biological processes expressed by American cellular biologist Lynn Margulis and science journalist Dorion Sagan: 'Life on Earth is more like a verb. It repairs, maintains, recreates, and outdoes itself'.[84] Margulis and Sagan borrow the term autopoiesis, defined as the process of self-making, from the second-order systems theory of Humberto Maturana and Francisco Varela, to trace the evolutionary formation of globally interconnected systems. These systems are symbiotic; they evolve through exchanging and acquiring genomes. Furthermore, Margulis and Sagan argue that

> we are only a single theme of the orchestrated lifeform. With its glorious nonhuman past and its uncertain but provocative future, this life, our life, is embedded now, as it always has been, in the rest of Earth's sentient symphony.[85]

As they explain, the inspiration for their 'sentient symphony' comes from Darwin's famous passage in the *Origin of Species*, describing life as 'an entangled bank' consisting of 'elaborately constructed forms, so different from each other, and dependent on each other in so complex a manner'.[86]

Similarly, London's reading of Darwin guides his interrogation of the perceived divide between animals and humans. For Darwin questions not just physical but mental differences. In *The Descent of Man*, he writes,

> [T]he difference in mind between man and the higher animals, great as it is, certainly is one of degree and not of kind. We have seen that the senses and intuitions, the various emotions and faculties, such as love, memory, attention, curiosity, imitation, reason, &c., of which man boasts, may be found in an incipient, or even sometimes in a well-developed condition, in the lower animals. They are also capable of some inherited improvement, as we see in the domestic dog compared with the wolf or jackal.[87]

London's dog narratives restage Darwin's ideas, speculating that evolution works equally upon and connects intelligence and physical strength as indicators of an organism's fitness. Some disagreed; most noticeably, in an interview published June 1907 in *Everybody's Magazine*, Theodore Roosevelt called London a 'nature-faker'. Roosevelt's accusation, however, was based on the incorrect claim that, in *White Fang*, a lynx kills a wolf. Pointing out this error, London's response, 'The Other Animals', published a year later in *Collier's*, defends his depiction of nature. He even admits that his concerns for scientific accuracy conflicted with standards for good writing:

> The writing of these two stories, on my part, was in truth a protest against the 'humanizing' of animals . . . I wrote, speaking of my

> dog-heroes: 'He did not think these things; he merely did them', etc. And I did this repeatedly, to the clogging of my narrative and in violation of my artistic canons; and I did it in order to hammer into the average human understanding that these dog-heroes of mine were not directed by abstract reasoning, but by instinct, sensation, and emotion, and by simple reasoning. Also, I endeavored to make my stories in line with the facts of evolution; I hewed them to the mark set by scientific research.[88]

This passage is worth quoting at length because it provides a snapshot of London's approach: dramatizing the theories of evolution that he read through documentary-like accounts of dog behaviour.[89] London goes to considerable lengths to deny dogs the ability to reason abstractly. Discussing Buck's moral regression, he explains: 'Not that Buck reasoned it out. He was fit, that was all, and unconsciously he accommodated himself to the new mode of life'.[90] London describes White Fang's perception in a similar way: 'Another advantage he possessed was that of correctly judging time and distance. Not that he did this consciously, however. He did not calculate such things. It was automatic'.[91] And although London addresses the shaping of human–dog relationships by the concept of justice, he comments that White Fang 'had no abstract ideas about justice and fair play'.[92] Overall London's 'narrative clogging' demonstrates his self-conscious working out of how Darwin's theories limit canine thought processes.

London's most visionary idea is that humans and dogs co-evolved. In *The Call of the Wild*, Buck's dreams about his wolf ancestors trace the biological kinship between humans and dogs:

> Sometimes as he crouched there, blinking dreamily at the flames, it seemed that the flames were of another fire, and that as he crouched by this other fire he saw another and different man from the half-breed cook before him. This other man was shorter of leg and longer of arm, with muscles that were stringy and knotty rather than rounded and swelling . . . He uttered strange sounds, and seemed very much afraid of the darkness.[93]

Building upon Darwin's contention that animals and humans have a common descent, London makes his case for biological kinship. Donna J. Haraway sheds light on London's argument by examining the idea of kinship as a non-objective 'technology for producing the material and semiotic effect of natural relationship, of shared kind'.[94] London's attention to the training and behaviours of working dogs pressures the idea of a universalized species discourse. Haraway's work on dog genetics provides an important perspective through which to evaluate London's depiction of the evolutionary links among wolves, dogs and humans. She explores the self-domestication of wolves that gradually produces a dog:

> Many interpreters argue that the most likely scenario has wolf wannabe dogs first taking advantage of the calorie bonanzas provided by humans' waste dumps. By their opportunistic moves, those emergent dogs would be behaviorally and ultimately genetically adapted for reduced tolerance distances, less hair-trigger flight, puppy developmental timing with longer windows for cross-species socialization, and more confident parallel occupation of areas also occupied by dangerous humans.[95]

Consequently, White Fang's domestication follows a historical and biological pattern: 'The qualities in his kind that in the beginning made it possible for them to come into the fires of men were qualities capable of development'.[96] London shows this development from various angles, from canine longings for 'the protection and companionship of man', to White Fang's 'own choice' to come 'to sit by man's fire and be ruled by him'.[97] Indeed, London anticipates what Haraway calls a capacity for 'cross-species socialization' in his description of White Fang's 'faithfulness': 'the quality that was peculiarly the possession of his kind; the quality that set apart his species from all other species; the quality that had enabled the wolf and the dog to come in from the open and be the companions of man'.[98] In 'The Other Animals', London punctuates his defence of co-evolution with the rejoinder, 'Let us be very humble. We who are so very human are very animal'.[99] His sentiments anticipate Margulis and Sagan's vision of a 'sentient symphony', of globally connected, autopoietic systems that refute human dominance.

What London seems to realize is that co-evolutionary processes that join animal and human also redefine the idea of vision. Maturana and Varela explain that because an autopoietic system is 'operationally closed' off from its environment, it 'does not "pick up information" from the environment, as we often hear. On the contrary, it brings forth a world by specifying what patterns of the environment are perturbations and what changes trigger them in the organism'.[100] The perception of dogs, then, is as subjective as that of humans, constituted through a field of vision that 'brings forth a world' and incorporates them in the construction of meaning. Preceding Buck, London writes,

> were the shades of all manners of dogs, half-wolves and wild wolves . . . scenting the wind with him, listening with him and telling him the sounds made by the wild life in the forest, dictating his moods, directing his actions, lying down to sleep with him when he lay down, and dreaming with him and beyond him and becoming themselves the stuff of his dreams.[101]

In *White Fang*, London connects subjective vision to a received set of experiences: 'Not alone out of his own eyes, but out of the eyes of all his ancestors was the cub now looking upon man – out of eyes that had circled in the darkness of countless winter camp-fires'.[102] Rather than an absolute perspective from which to observe animal thinking, London gives us partial glimpses of evolution in action and identities in formation. As Haraway comments, gender '[d]ifference is theorized *biologically* as situational, not intrinsic, at every level from gene to foraging pattern, thereby fundamentally changing the biological politics of the body'.[103] At the same time, if domestication is a process that remains open to environmental changes and animal agency, the idea of a civilized endpoint as class marker, and race (and, by extension, gender) as a means of naming cross-species relationships, is disrupted. What appears as London's most imaginative yet paradoxical deconstruction of evolutionary being is, at the same time, a reconstruction. That is, the aforementioned image of dogs 'dreaming with [Buck] and beyond him and becoming themselves the stuff of his dreams' dissipates an essential self for Buck. London's paradoxical way of questioning a static model of subjectivity, moreover, fits with the paradoxical constitution of autopoietic systems. As Wolfe's reading of systems theorist Niklas Luhmann elaborates,

> Luhmann insists that all observations . . . are based on systems codes, which are themselves built on a constitutive distinction that is essentially paradoxical or tautological. What this means . . . is that all such observations and systems are built on an inescapable 'blind spot' that only *other* observations, from within *other* systems can reveal.[104]

London mirrors this process by the generations of dogs, half-wolves, and wolves that constitute Buck and yet act apart and 'beyond him'. In the words of Luhmann, for a system, 'the condition of its possibility is its impossibility'.[105] London's self-reflexive depiction of the evolutionary process presents the set of possibilities for a subjectivity in becoming that reconstructs gender, race and class markers used to delimit an individual self.

While London's dog narratives try, at various times, to construct gender, race and class as anchors for subjectivity, the slipperiness of these categories emerges in the images I have been addressing. The images turn subject making inside out by questioning the boundaries of corporeal singularity. In the images, Buck and White Fang appear as multiple assemblages of dog bodies, displaying what Jacques Derrida would later recommend:

> to have the plural of animals heard in the singular. There is no animal in the general singular, separated from man by a single indivisible limit. We have to envisage the existence of 'living creatures' whose plurality cannot be assembled within the single figure of an animality that is simply opposed to humanity.[106]

London's dogs undo the monolithic, human-centred construction of gender, race and class that can be articulated only through notions of singularity. In this traditional view, as Wolfe interprets Derrida's argument against it, 'the Word, *logos*, does violence to the heterogeneous multiplicity of the living world by reconstituting it under the sign of identity, the *as such* and *in general* – not "animals" but "*the* animal"'.[107] In the images of Buck and White Fang, agency exceeds a 'sign of identity'. To put it another way, what an animal *does* both can and cannot determine what an animal *is*. The self can *only* be seen through a performativity that is *not* self-identical; it goes beyond the limits of (self-)closure.

For London, dogs are therefore in a constant process of 'becoming' rather than fixed biological types. They evolve in different ways according to autopoiesis, a process guided by environment and heredity. In Buck's case, the 'heredity of the breed' helps him to adapt to the environment, by giving him the atavistic traits, such as learning to 'fight with cut and slash and the quick wolf snap', needed for survival in the Northland.[108] In *White Fang*, the process of becoming disrupts a secure, self-identical subjectivity. Demonstrating London's belief that '[e]very atom of life is plastic',[109] there is a confusion and conflation of identities that constitute White Fang:

> His heredity was a life-stuff that may be likened to clay. It possessed many possibilities, was capable of being moulded into many different forms. Environment served to model the clay, to give it a particular form. Thus, had White Fang never come in to the fires of man, the Wild would have moulded him into a true wolf. But the gods had given him a different environment, and he was moulded into a dog that was rather wolfish, but that was a dog and not a wolf.[110]

London stages the same environmental and hereditary conditions for the domestication of dogs that Haraway describes. We are reminded, moreover, that the domestication of White Fang is an ongoing process, 'a matter of development'.[111] Overall then, London's vision of biological kinship destabilizes human subjectivity, something that human reason, always and already guided by evolution, cannot refute. As he concludes in 'The Other Animals',

> You must not deny your relatives, the other animals. Their history is your history, and if you kick them to the bottom of the abyss you go yourself. By them you stand or fall. What you repudiate in them you repudiate in yourself – a pretty spectacle, truly, of an exalted animal striving to disown the stuff of life out of which it is made, striving by use of the very reason that was developed by evolution to deny the process of evolution that developed it.[112]

But London's views on biological kinship rely on claims about the mental similarity of higher animals to '[t]he lowest human types' who 'do little to no reasoning in the abstract'.[113] A cultural model of gender, race and class differences is installed at the heart of his conception of biological similarity. In the end, London can reformulate the categories of animal and human only at the cost of making some humans *less* than human, that is, by imagining that there is a capability, abstract reasoning, that not all humans possess.

IV. Ghosts of American citizens

A notable problem with London's schematic for biological kinship appears through the way in which his dog narratives are subsumed in the biological politics of the frontier, articulated through discourses about gender, race, class and modernization. These discourses animate US imperialist desires through fantasies of masculine violence that secure and maintain a national identity. According to Annette Kolodny, the linking of violence to subject making indicates that 'the new nation had entered its adolescence' and created 'a field for exercising sexual mastery and assertive independence, its conquering hero captured in the legendary figure of John Henry, the steel-driving man'.[114] The setting of both novels in the Yukon, during the gold rush, draws attention to the control and domination of the natural landscape. In *The Call of the Wild*, Buck's winning the bet funds Thornton's 'trip into the virgin East, where miners had not yet appeared'.[115] The image of penetrating the 'virgin' land to mine gold is undeniably sexual, a well-rehearsed image of taking a feminized nature by masculine force. It also reminds us that London's scientific documentary of dogs carries with it undertones of the violation of nature done in the name of science. As Evelyn Fox Keller points out: 'The ferreting out of Nature's secrets, understood as the illumination of female interior, or the tearing of Nature's veil, may be seen as expressing one of the most unembarrassedly stereotypic impulses of the scientific project'.[116] In a parallel to Buck's evolutionary regression, Thornton's search for gold restages the history of the exploitation of the natural landscape. The conversation, for instance, that initiates the bet occurs in the Eldorado Saloon, referencing, as Michael L. Johnson notes, the mystical city of gold sought by Spanish explorers and starting a pattern of at best ignoring and at worst despoiling the natural beauty of the landscape.[117] And in practice, gold-mining, as Johnson details, made 'sludge and rubble of geographical splendors' as well as effecting 'urban growth and the overnight blooming of boomtowns that often subsided into ghost towns'.[118] In London's novel the town is full of sudden wealth that produces a nouveau riche class, embodied by 'a Bonanza King' and 'a member of the latest dynasty, a king of the Snookum Benches' who represent the capitalist taming of wild lands.[119] Furthermore, in Auerbach's view, the bet on Buck reflects the economic drive that exploits both land and animals for profit.[120] Thornton's quest connects past abuses of the land with the present desire for US expansion and economic progress. He searches for a contemporary equivalent of Eldorado, 'a fabled lost mine, the history of which was as old as the history of the country'. Roosevelt's promotion of the 'strenuous life' informs the portrayal of the search as nature's test of masculine strength, 'to achieve where men and dogs as good as themselves had failed'.[121] At one point, Thornton finds 'a Hudson Bay Company gun of the young days in the Northwest, when such a gun was worth its height in beaver skins packed flat'.[122] A historical symbol of masculine dominance, the gun recalls the history of capitalist trade in animals that makes the land a profitable commodity.[123]

While Thornton and his party do eventually become rich, Buck communes with the wild, spending ever more time away from the camp and undergoing the final stages of atavism. The relationship between Buck and Thornton is disrupted, but not before it has been described, as in *White Fang*, as a heterosexual arrangement. This time, however, Thornton acts the female role. Previously, he had been the object of Buck's male gaze.[124] Auerbach finds that at the camp, Thornton's 'female domesticity' and his proclivity for 'sentiment and tears' cause him to resemble a 'wronged' woman, who must 'wait at home for the straying, unfaithful lover now "seized" by irresistable impulses . . . and wild yearnings'.[125] Yet one of Buck's last displays of affection outdoes even the antics of Skeet or Nig: 'John Thornton was eating dinner when Buck dashed into camp and sprang upon him in a frenzy of affection, overturning him, scrambling upon him, licking his face, biting his hand – "playing the general tom-fool," as John Thornton characterized it'.[126] We should take careful note of the description of Buck 'playing' a more feminine role. In its sheer flamboyance, this resembles a canine performance in drag. Rather than subvert dominant gender roles, however, Buck's cross-gender acting only accentuates the male performing that follows it, just as Sandow's US stage debut as the perfect man came after a performance of the farce *Adonis* (a gender-reversed version of the Pygmalion story in which a man is pursued by women).[127] In a virtual cataloguing of Sandow's physical traits, Buck's body is 'magnetic' when touched, and '[e]very part, brain and body, nerve tissue and fibre was keyed to the most exquisite pitch; and between all the parts there was a perfect equilibrium or adjustment'.[128] If Thornton has the final word – 'Never was there such a dog' – then Buck exhibits how striking poses and being looked at constitute and maintain the iconography of masculinity.

Such is the case with Sandow, whose frequent pose as gladiator, Kasson explains, is significant because it relates to a 'growing celebration of America's historic encounter with the primitive', epitomized by Turner's famous speech, 'The Significance of the Frontier in American

History', given in Chicago in July 1893, 'less than three weeks before' Sandow's premiere in the same city. According to Kasson, Turner celebrated the frontier, even though it had already closed in 1890, as 'the historic site not only of American individualism and democracy but also (though he did not make this claim explicit) of American manhood'.[129] Displaying the roots of manhood in atavistic impulses, Sandow's gladiator pose fits into a larger doctrine of 'self-making' that inspired boys from all class backgrounds. One of these boys from the upper class, Roosevelt, would help to reinvent the frontier, an act that guides Buck's final transformation.[130]

After killing a moose, his initiatory rite into the wild, Buck revisits the camp just in time to avenge the murder of Thornton, his comrades and dogs by a Native American tribe, the Yeehats. Such a problematic narrative turn – why Buck should *return* from the wild – can be explained through Slotkin's frontier thesis. He writes that the frontier, as an ideological construct, 'represented the redemption of American spirit or fortune as something to be achieved by playing through a scenario of separation, temporary regression to a more primitive or "natural" state, and *regeneration through violence*'. Buck's evolutionary regression, signified by his identification with the untamed wolf rather than domesticated dog, recaptures a masculine vitality expressed through his attack on the Yeehats. As Slotkin explains, 'regeneration through violence' is effected through a 'myth of "savage war"' that 'blames Native Americans as instigators of a war of extermination'. The Yeehats' attack by stealth on Thornton's party thus justifies Buck's slaughter. His vengeance, in historical context, not only legitimates US dominance within its borders but rationalizes national expansion into new frontiers. A similar form of rationalization, Roosevelt's speech 'The Strenuous Life', as Slotkin explains, connects the 'virility' of economic leadership to 'the challenge of empire'. Roosevelt transforms the symbol of the 'Indian frontier' into 'a metaphoric extension of Frontier categories to a new situation in which Asians become figurative Apaches and the Philippines become a symbolic equivalent of [Daniel] Boone's Kentucky or [Sam] Houston's Texas'.[131] In other words, the image of the Native American vanishes from the United States and reappears in lands ripe for the imperialist taking, as if to set the stage for conquest. The restructuring effect of imperialist strategies on cultural formations of race, as Shelley Streeby's discussion of the 1846–8 US–Mexican War illustrates, extends into the early twentieth century with the depiction of Native Americans.[132] The reinvention of the US winning the frontier displaces the ambivalence about the 'vanishing Indian': the feeling that, valued for their apparent primitive virtues, Native Americans could not survive in a modernized present, and that they were therefore 'doomed to extinction' as victims of civilized progress.[133] Expressing the ambiguity of the value of a 'wild' masculinity threatened by capitalism, London seeks a solution through escape; at the end of the novel, Buck, as a 'Ghost Dog', becomes incorporated into the 'frontier myth' of regeneration through violence, eluding all social constraints.[134] The male iconography in the novel thus shifts from the perfect man, which supported scenarios of racial violence, to the escape artist, mirroring how Harry Houdini surpassed Sandow in turning male performances into spectacles for a mass audience.[135] Formerly Ehrich Weiss, he first escaped his Jewish immigrant origins, renaming and remaking himself as Houdini.[136] Throughout, Kasson elaborates, his performances traversed gender roles; in one trick performed in the 1890s, he exchanged places with his wife, who then appeared bound, in a reconfirmation of male dominance versus female submission.[137] In a trick in the early 1900s, 'he placed himself in what might be regarded as an especially provocative, feminized position: naked, bound, bent over', giving the initial impression of being 'trapped in a position of humiliating gender defeat – unmanned – only to achieve a grandiose masculine triumph'.[138]

Yet if Buck, as canine escape artist, appears to transcend the cultural boundaries of gender, Kasson reminds us that 'Houdini cultivated illusion in order to reaffirm traditional verities of masculinity'.[139] Hence traditional gender subjectivity relies on the illusion that gender is stable and fixed. To maintain traditional gender roles, however, these roles must remain in a state of reconstruction.

Furthermore, the ghost dog displays how national subjectivity is shaped by the uncanny. Priscilla Wald's study of Sigmund Freud's use of the uncanny locates the concept (from *unheimlich*, 'not homelike') within the production of 'official stories of the nation'.[140] The uncanny 'turns on the discovery that the unfamiliar is really familiar (the stranger as self) . . . [It] sends us home to the discovery that "home" is not what or where we think it is and that we, by extension, are not who or what we think we are'.[141] The phantoms of race haunt London's vision of the frontier as a site for national subjectivity; that the frontier is a liminal space between culture and nature echoes the legal distinctions between citizens and non-citizens. Supreme Court decisions on *Cherokee Nation v. the State of Georgia* (1831) and *Scott v. Sanford, or*

Dred Scott (1857), Wald explains, placed the identities of indigenous Native Americans and African descendants in an ambiguous space between 'citizens' and 'aliens'.[142] A similar decision was made concerning the identity of Puerto Ricans in *Downes v. Bidwell* (1901). These cases collectively suggest that 'the emerging imperial nation once again legislated its ghosts into existence in the name of the (white) American family, and once again these presences were haunted by virtue of their resemblances as forcefully as their differences'.[143] Buck, as ghost dog, reflects the same anxieties about the vanishing Indian expressed at the turn of the twentieth century by Barrett Wendell, a Harvard English professor, who writes: 'The very name of us means something not ourselves . . . I feel as I should think an Indian might feel, in the face of ourselves that were'.[144] Fantasies of frontier conquest, in *The Call of the Wild*, compensate for but do not alleviate the imperial homelessness caused by a national subject always in the process of becoming, but never arriving.

In *White Fang*, the idea that subjectivity remains ambiguous is reflected in the loss of the natural wildness of the landscape. As in *The Call of the Wild*, mining is connected to capitalist enterprise, this time through White Fang's encounters with white men, seen (by the author, by dominant culture and by the dog, who projects the author and his culture's racial views) as a superior race. With Gray Beaver, he arrives at the Yukon, the gateway to the gold-rich territories of Dawson and the Klondike, where 'stood the old Hudson's Bay Company fort . . . It was at Fort Yukon that White Fang saw his first white men. As compared with the Indians he had known, they were to him another race of beings, a race of superior gods'. London associates the historical conquest of the land, requiring military outposts to protect capitalist trade, with racial superiority. Economic and racialist discourses are further naturalized through White Fang's perception. Although White Fang endorses the belief that white men are naturally superior, he 'does not reason it out'; it is a 'feeling'.[145] The narrative trajectory reinforces the idea that biological kinship is tied to capitalist social networks, for White Fang is saved from death by Scott, a mining expert who has ties with the Gold Commissioner.[146] As a mining expert, Scott possesses the technical knowledge and scientific mastery for extracting resources from the land and initiating the transformation of wild lands into civilized regions. Representing a new managerial class of men in the Roosevelt era, Scott embodies an economic progress that constitutes, in Slotkin's words, 'a newly empowered elite', just as White

Fang embodies the evolutionary progress that produces a hybrid of wolf-dog.[147] The image of the Hudson Bay Company, which also appears in *The Call of the Wild*, frames the making of White Fang into Scott's domesticated companion animal. Regarded as companion species, dogs, Haraway reminds us, historically served as 'lethal guided weapons and instruments of terror in the European conquest of the Americas'.[148] In London's narrative, fantasies of conquest and national economic progress inform the biotechnological apparatus that connects humans and dogs through environmental exploitation. The novel thus invokes a regime of what Greta Gaard calls 'mutually reinforcing systems of oppression' currently active in economically disenfranchised global regions.[149] Indeed, the relationship between the Southland and Northland reflects, albeit subtly, how imperialist policies justify the depletion of labour and resources from these regions.

The civilizing of White Fang in the Southland is juxtaposed against the threat of cultural primitivism, embodied not by Native Americans but by a different cultural other, Jim Hall, a 'human beast' who is not naturally formed but rather created by 'the moulding he had received at the hands of society'.[150] Hall escapes prison and traverses a visibly animalistic landscape that reflects his identity as a primitive brute, pursued by 'the sleuth-hounds of the law, the paid fighting animals of society'.[151] Attempting to exact revenge on the judge, Scott's father, who had sentenced him, he breaks into the house. While the narrator expresses some sympathy for Hall, who was 'innocent of the crime for which he was sentenced', it is implied that Hall is a threat to the masculine order that must be controlled, violently if necessary.[152] True to his training to uphold patriarchal law, White Fang waits until Hall starts up the stairs that 'led to the love-master and the love-master's dearest possessions' before he attacks.[153] His killing of Hall protects Scott's father, who, as a judge, embodies the authority of patriarchal law. This scene of masculine violence recovers traditional gender roles: while White Fang recovers from being shot by Hall, the women of the family play the role of nurses. Although he struggles for his life, his natural strength prevails, for 'a constitution of iron and the vitality of the Wild were White Fang's inheritance'. What is transmitted biologically is the inheritance of whiteness, leading to and preserving social order. His recovery complete, his masculine subjectivity is re-established when the Judge names him as 'Blessed Wolf'; and White Fang in turn becomes a father himself, thus reproducing the patriarchal values that he has internalized.[154]

A nightmare that White Fang has near the end of the novel, however, again suggests the uncanny at work; here a repressed wildness returns in the form of a technological monster. In his nightmare, a modernized, mechanized society depletes the masculine 'vitality of the Wild': a squirrel 'transform[s] itself into an electric car, menacing and terrible, towering over him like a mountain, screaming and clanging and spitting fire at him'.[155] Like the anxieties of Wendell, the dream reveals the haunting effects of the erasure of Native Americans in national culture. The biological taming of wild territories reinforces racial hierarchies, unfolded in the narrative of national economic progress, yet constrains individual agency. Before he is named by the Judge, White Fang already must deal with a patriarchal restructuring of the cross-species relationships that shape him as an animal agent: 'In the end he worked out the complete law. Between him and all domestic animals there must be no hostilities . . . But the other animals – the squirrels, and quail, and cottontails, were creatures of the Wild who had never yielded allegiance to man'.[156] What he is permitted to kill marks the patriarchal law that exerts control over the natural world, an arrangement reiterated by the act of his naming. Again, the performativity of naming, that is, bringing into being in a legal sense what is named, shows according to Butler how 'it conceals or dissimulates the conventions of which it is a repetition'.[157] Under the guise of natural adaptation, human dominance over nature is justified through the narrative of White Fang's development. And Auerbach observes: 'Taming exacts its price: although White Fang may show some of his old spirit in the end by saving his adopted masters from an escaped prisoner bent on murder, the wolf-dog remains a defanged family pet more akin to Lassie than Buck, despite London's best efforts to retain some of the beast's wildness'.[158] The final scene stages Butler's speculation that 'if there is agency, it is to be found, paradoxically, in the possibilities opened up in and by that constrained appropriation of the regulatory law, by the materialization of that law, the compulsory appropriation and identification with those normative demands'.[159] If Butler theorizes, moreover, that 'becoming is no simple or continuous affair, but an uneasy practice of repetition and its risks', then a repeated risk expressed in London's novels is the anxiety, as Wendell laments, that we can never be truly ourselves.[160] As we will see, legal attempts to secure a national subjectivity destabilized by evolution and imperial conquest encounter similar problems.

V. Where to draw the line? Biological kinship and legal discourse

London's 23 June 1899 letter to Johns rehearses the argument he will make in 'The Other Animals':

> Do you know that the physiologists say that the difference between the highest forms of man and the lowest forms of man is greater than the difference between the lowest forms of man and the highest forms of the rest of the vertrebates [*sic*]? This being so (and it is so), where are you to draw the line? If the lowest forms of man are to be preserved, no matter what the moral logic involved, then, by that same logic, you must not exclude the higher forms of the rest of the vertrebates [*sic*].[161]

Despite the fact that 'higher' and 'lower' immediately construct an evolutionary hierarchy, human and non-human animal subjects remain fluid. London's language is not exclusionary; it takes into account the inability to represent a fixed definition of the difference between animals and humans. But a subsequent letter to Johns, dated 5 June 1899, uses the idea of a natural law 'to draw the line' and, in doing so, reintroduces exclusionary language into his concept of biological kinship:

> [W]e may come to know the law of some of the forces . . . These forces generated the altruistic in man; the race with the highest altruism will endure – the highest altruism considered from the standpoint of merciless natural law, which never concedes or alters. The lesser breeds cannot endure.[162]

London's tendency in the novels to conflate natural and legal law, to use exclusionary language from legal discourse to regulate the fluid subjects produced by biological kinship, extends to larger attempts to fashion a universal code for rights, based upon a similar legal language. In the same year as *White Fang*, John Howard Moore's *Universal Kinship* was published. In his personal copy, London marked a passage that resonates with his thinking: '*All* beings are *ends*; *no* creatures are *means*. All beings have not equal rights, neither have all men; but *all have rights*'.[163] *The Call of the Wild* and *White Fang* demonstrate, however, that legal appeals for subject agency are haunted by the ambiguity over racial citizenship. And, while the Southland and Northland appear to be oppositional in nature, in reality they are both shaped by an upper-class managerial ideology that supports legal claims for mining (and maintains ownership over natural lands), justifies the unequal social status of the Northland and Southland and parcels out global regions for imperial coercion.

In this light, it is worth looking at Wolfe's commentary about the blind spots inherent in autopoietic systems. He explains:

> In second-order observation we are observing observations – and observing, moreover, how those observations are constructed atop a blindness to the wholly contingent nature of their constitutive distinction. (The legal system, for example, cannot carry out the observations of legal versus illegal while at the same time recognizing the essential identity of both sides of the distinction, its essential tautology: legal is legal.)[164]

Like the legal system, which Wolfe points out must function without being able to recognize essentially what is not legal, London's legal appeal for subject agency cannot recognize what biological autopoiesis does not validate: animals as a singular (not multiple) category, a monolithic natural law and higher and lower human subjects. His reading of evolutionary theory as supporting a human-centred and race-based notion of altruism thus reinforces the ecological separation between nature and culture, between human and animal, that animates the empire building of the early twentieth century.

4

AFTER THE FLOOD: PERFORMANCE AND NATION

In the early twentieth century, a critical ecological issue was water purification. Published in December 1900 in *Popular Science Monthly*, George C. Whipple's 'Municipal water-works laboratories' documents the emerging policy of biopolitical control of life (a concept which will be discussed later). Read through Whipple's article, this policy results from the scientific comingling of modernity, nationalism and pragmatism:

> The laboratory idea is fast taking hold of our municipalities. It is the natural result of modern science and American practicality. More and more our civilization is making use of the great forces of nature, and more and more it is becoming necessary that nature's laws should be understood: hence the need for the precise data of the expert and the long-continued observations of the specialist. This is emphatically true in the domain of sanitary science, where the advances in chemistry, microscopy, and biology have wrought revolutionary changes. The microscope is no longer a toy, it is a tool; the microscopic world is no longer a world apart, it is vitally connected with our own. The acceptance of the germ-theory of disease has placed new responsibilities upon health authorities and has at the same time indicated the measures necessary to be taken for the protection of the public health.[1]

Whipple foregrounds the act of observation by associating it with specialized interests, its importance reinforced by the image of the microscope that connects human and non-human worlds. Yet the primacy of an objective point of view is troubled, because the microscope reminds us of the technologically mediated quality of the observations of a world that cannot be accessed by *human* vision alone. We therefore arrive at a transition from first- to second-order observation.

While Whipple remarks that 'supplying water to a community' becomes 'an engineering problem',[2] such a technical objective is achieved through scientific experimentation with different forms of filtration. In particular, different systems of mechanical water filtration are attached to nationalistic signifiers: in the prime example, the 'American system' differs from the 'English system', the latter using natural sand filtration.[3] Experiments on the Ohio River from 1898 to 1899 confirmed that the 'American system could be operated with less difficulty and with somewhat less expense'.[4] The economic language of cost efficiency figures into the ways that ecology is nationalized, suggesting a reaffirmation of national control over natural resources. Whipple, moreover, promotes the value of scientific experimentation in the global marketplace, for the findings of such experiments 'have been not only of local and immediate value, but they have acquired a world-wide reputation and form a permanent contribution to scientific literature'.[5]

The conclusion of the article resonates with Edith Wharton's *The Custom of the Country* (1913). Whipple outlines a centralized model of biopolitical control:

> It seems apparent, therefore, that the laboratory is destined to be an important factor in municipal engineering as well as in municipal sanitation, and it is not difficult to foresee a time when every city of importance will be provided with a laboratory equipped in accordance with its needs. In large cities, work of this kind is preferably specialized and distributed through different departments, in order that it may be under control of those directly interested in the results, but in small cities, all the analytical work can be more economically accomplished in a single laboratory. In such a laboratory the work would cover a very broad field . . . All of this would require the services of an engineer, a chemist and a bacteriologist, or of these three combined in one person. The expense of such an institution would be small in comparison with the saving that would result to the citizens in the purchase of supplies and in the protection of the public health.[6]

We can see why water management is a crucial part of the capitalist agenda described in Wharton's novel. To hold a dominant share of a significant resource, one that applies for all US cities – but especially those in post-frontier zones – only requires exerting influence at the top of the management hierarchy, and in such a centralized model, all 'expenses' could be written off as necessary for 'the protection of the public health'. Of course, what happens ethically when competing interests vie for control of urban water resources is not easy to glimpse in Wharton's narrative because of the murky details of these large-scale, interstate transactions.

Wharton boils down these ethical concerns into depictions of secret water-deals and a femme fatale, anticipating by half a century Roman Polanski's *Chinatown* (1974), written by Robert Towne and starring Jack Nicholson and Faye Dunaway. The film, set in 1937, references the 1908 Los Angeles water scandal. Wharton's *The Custom of the Country* tells a similar story about a Midwestern city and how the social mobility of Undine Spragg, the protagonist, is enabled by her father's participation in the 'Pure Water Move'. *Chinatown*, with its sun-drenched landscape, reinvents the visual look of the film noir; Wharton's novel innovatively connects water regulation to a network of social regulation in visual media. Polanski's innovation stops at the local level; *The Custom of the Country*, however, goes beyond Polanski's invocation of environmental crisis by mapping international economies onto global ecosystems.

If the key line of the film is 'Forget it, Jake, it's Chinatown', Wharton's novel most definitely does not allow us the privilege of forgetting the mounting perils of corporate control over water resources which fits with a business ethics that rationalizes exploitation of global labour to establish the United States as lead performer on the world stage. The cost, I will suggest, is in terms of biopower (the making of government policy for the public benefit), and that cost becomes visible in a myriad of places, in particular an art studio in New York City.

I. Managing life

The 'Pure Water Move' in Apex City is a convoluted business transaction that is viewed somewhat differently by every character, and warrants a careful review. Indeed, Wharton embeds this non-linear ecological narrative in a framework that anticipates the environmental justice mysteries of the later twentieth century.[7] We first hear of the transaction through Ralph Marvell who is courting Undine Spragg. While the inspiration for her parents to name Undine comes 'not from the water sprite of classical myth, but a commercial hair-curling product',[8] she has family ties and, later, marriages with those who are complicit in turning water management into economic gain. As Leota, Undine's mother, explains to Ralph, the Pure Water Move is a heroic, transformative and redemptive act by her husband, Abner. In Leota's account, it is shaped by the daily struggle of living in a frontier town in the throes of modernization. After a typhoid epidemic kills two of his three children, sparing Undine, Abner pledges that 'Apex should drink pure water'.[9] Taking over land from his father-in-law for an unpaid debt, he makes the Pure Water Move. The land is then sold for the site of a new reservoir; in Leota's words, '[A]fter that, we began to be better off, and it *did* seem as if it had come out so to comfort us some about the children'.[10]

Ralph is unable to 'read between the lines of Mrs Spragg's untutored narrative'.[11] Although we never find out what *exactly* the Pure Water Move is, we learn later that Representative James Rolliver had given Abner crucial assistance, leading him 'through the muddiest reaches' of the transaction.[12] Abner becomes an overnight success in Apex City, which enables his family's relocation to New York City. But he has traded an environmental crusade against contaminated drinking water for an economic battle: to make it in the competitive Wall Street world, he must outsmart his rivals and continually invent and develop lucrative business deals, which he eventually proves unable to do.

Moving, moreover, does not completely free the Spraggs from the social and sexual scandal that hastens their departure from Apex City. Elmer Moffatt, whose freewheeling behaviour equals his ambition, also moves to New York City. Earlier, in Apex City, his reputation is badly damaged when he appears drunk at a Temperance Society ceremony; he then elopes with Undine, but their marriage is undone by Abner with the backing of Rolliver. Forced out of Apex City and his job at the water-works powerhouse, Moffatt goes to Alaska and somehow becomes involved in the Eubaw mines. A powerful financier prominently involved in the Ararat Trust, Harmon B. Driscoll, hires Moffatt, who has insider information, as one of his personal secretaries. With Moffatt's help, Driscoll acquires the mines. Driscoll also owns the street railroads of Apex City. He sets his sights on water resources and enlists Moffatt to persuade Abner to divulge what he knows about Rolliver, Driscoll's rival. As Moffatt explains to Abner during their meeting in New York City, there will be a financial reward for his help. Moffatt's offer, however, is a form of blackmail;[13] he knows about Undine's engagement to Ralph, and in exchange for Abner's cooperation, he will remain quiet about Undine's past. Abner accepts the offer, and the proceeds fund a rushed yet lavish wedding for Undine, who has promised Moffatt that she will help him enter her 'new circle'. But Abner's compliance has consequences, for Rolliver will continue to hold a grudge against him, ensuring that Abner will never attain financial stability.

Through discussions among Undine's circle, we hear about the investigation of the Ararat Trust, which Moffatt, taking advantage of the opportunity to damage Driscoll, actively supports. The investigation, however, is 'mysteriously stopped',[14] a serious setback for Moffatt. Clare Preston explains his recovery:

> Moffatt tries to interest Mr Spragg in a new real estate deal (presumably also connected with Apex), but Mr Spragg will have none of it. Moffatt then approaches Ralph through Undine, since Ralph's real estate firm is concerned in the transaction. It appears that Moffatt's scheme is the (characteristically) secret acquisition of a crucial piece of land which two major parties are angling for (possibly his own friends and enemies Driscoll and Rolliver); and with Ralph's help he gets the land. Thereafter we hear the Ararat investigation is being reopened, and that Moffatt is required as state's witness.[15]

The next instalment of the saga unfolds through Ralph's conversation with Clare Van Degen, his cousin, who has agreed to fund his legal battle with Undine over custody of his son, Paul.

Ralph tells Clare that he has invested her money in Moffatt's latest gambit, a collaboration with Rolliver who 'practically controls Apex': 'This is a scheme to buy up all the works of public utility at Apex. They're practically sure of their charter'.[16] Monitoring his investment, Ralph, 'from various sources, continued to pick up a good deal of more or less contradictory information',[17] including that Driscoll has bought Moffatt's silence in the second investigation, and the news of Moffatt's subsequent rise to financial and social prominence. That Ralph appears confused throughout about Moffatt's strategy, which extends to the reopening of the Ararat investigation, seems an integral part of the narrative, one intended to underline the complexity of the financial deals. While considerable delays instigated by reformers add to the confusion, Moffatt and Rolliver eventually get the charter; they drive out Driscoll, and Apex City becomes theirs. But the windfall comes too late to benefit Ralph, who, in a state of mental collapse, has killed himself.

Overall, the significance of the legal response, constituted as an inquiry into the business affairs of the Ararat Trust, is preceded and somewhat undermined by the social and sexual scandal that implicates Undine, the prototypical femme fatale, who is linked romantically to both Moffatt and Ralph – and illustrates the media's penchant for elevating the latter over the former, due in no small part to a preference for reporting the affairs of corporeal bodies over corporate bodies. Together, the investigation and scandal trace out Michel Foucault's notion of biopower, in which

> [M]ethods of power and knowledge assumed responsibility for the life processes and undertook to control and modify them. Western man was gradually learning what it meant to be a living species in a living world, to have a body, conditions of existence, probabilities of life, an individual and collective welfare, forces that could be modified, and a space in which they could be distributed in an optimal manner. For the first time in history, no doubt, biological existence was reflected in political existence…[18]

From Foucault's writing about the eighteenth century, we obtain the larger and critical idea that the emergence of bio/political systems makes sexual reproduction integral to the 'conditions of existence'. As a signifier of corporeality, sex becomes crucial to biopower, due to an emphasis on life and the practice of living. Foucault observes that 'at the juncture of the "body" and the "population", sex became a crucial target of a power organized around the management of life'.[19] As a determinant of social well-being, biopower, articulated as how to guarantee public access to potable water, touches on foremost issues of morality. Yet, as Cary Wolfe elaborates, the connections between biopower and morality are complex. To begin with, there is certainly a legal apparatus (a legal precedent, even) behind the Pure Water Move that shifts into what Foucault calls a 'regulatory' role.[20] Wolfe's study of bioethics, the most current version of biopower, however, questions whether any model of regulation can carry out anything like a morally guided programme: if bioethics serves as the 'self-designated conscience' of 'institutions of state and economic power', then there is a fundamental disjuncture between 'the functions of conscience and those of establishing policies palatable to both state and economic power'.[21] We can see that such a disjuncture has been there from the start, as envisioned by Wharton's novel. In the lengthy narrative, where the Pure Water Move (as the epitome of 'state and economic power') becomes an underlying force that shapes the stages for social performance, we seem to find little public concern about the moral issues involved in treating waterworks (an essential public service) as nothing more than bargaining chips for large-scale, interstate, economic negotiations. Yet the name of the Ararat Trust, invoking the mountaintop where Noah's Ark landed after the flood, suggests an apocalyptic vision of environmental catastrophe that unsettles the mechanistic doctrine of what Mark Seltzer calls the 'channelling of natural floods into orderly movements' necessary for public water management.[22] In this light, there is an ironic dimension to Moffatt's offhand reference to 'the Day of Judgment',[23] denoting when the charter will be finalized. Even at the local level of the story, we should note Leota's explanation – her defence of the family's economic well-being from the 'shady' transaction that may have accelerated the family's departure from Apex City[24] – which in her view mitigates the sadness of the children's deaths (to be sure a *curious* view, overall, of morality).

Thus as Wolfe quotes from bioethicist Carl Elliott, the idea that '[o]ur moral attitudes are not *grounded* by a theory of persons; they are *built into our language*' pressures how moral attitudes 'can be distinguished from a simple appeal to social norms and, more pointedly, the prejudices and pragmatic expediencies on which such consensus is based'.[25] And in *The Custom of the Country*, we get an idea of how murky the associations are between moral attitudes and social norms. While Abner is outraged at Undine's desire for a life distinctly outside the social boundaries of her marriage, his reaction is qualified through the narrator's comment that his 'private rule of conduct was as simple as his business morality was complicated'.[26] About the backroom deals involving Abner that created the Pure Water Move we can only guess; later, Undine hears through 'snatches of smoking-room talk' about Moffatt's participation in the first Ararat Trust investigation, which interests her in so much as she is still attracted to Moffatt, and that the investigation has put the Driscoll family in financial straits, thus threatening the cancellation of their 'fancy ball'.[27]

It is, however, Peter Van Degen's remarks that redefine the notion of biopower in the novel. An upper-class playboy who shamelessly pursues Undine, Peter praises Claud Popple, a portrait painter favoured by Peter's class because he privileges 'elegance' over 'art'.[28] After viewing Popple's portrait of Undine, Peter enthuses that '*he* knows how we live and what we want',[29] restating biopower ('how we live') and ethics as business morality a justification of 'what we want'. As Nancy Bentley observes, Peter's words trace 'new paths of desire, social energy, technology, and capital . . . through a series of interconnected sites of public representation'.[30] While Bentley focuses on how Peter's words suggest 'the astonishing movements of Undine', something on which I will comment in a moment, it is worth considering how Undine dramatizes the transformation of the frontier, a significant social change. Her 'pioneer blood', a corporeal image that condenses the complex politics of biopower in the Western frontier, signifies her membership of a new social class (the nouveau riche) that threatens patricians such as Ralph.[31] But the image is displayed at a time when the referent (the frontier) for her identity has shifted. Due to the closing of the frontier in 1890 (as discussed in Frederick Jackson Turner's thesis), Apex City (modelled on Kansas City), defined by Richard W. Etulain as a Western city, is situated in a 'post-frontier region'.[32] The post-frontier signals a loss of the sense of place that anchors national identity. Witness Raymond de Chelles's criticism, which equates Undine's restless lifestyle with a larger national state of disorder: '[Y]ou come from hotels as big as towns, and from towns as flimsy as paper, where the streets haven't had time to be named, and the buildings are demolished before they're dry'.[33]

Ralph's evolutionary fantasy fits with the end of the frontier. He calls his own family 'Aborigines', 'those vanishing denizens of the American continent doomed to rapid extinction with the advance of the invading race'.[34] It is no mere coincidence then that Undine's divorce attorneys are located in Sioux Falls, South Dakota, a region whose name suggests a 'vanishing' Native American presence.[35] By trying to 'renaturalize' the frontier as a site for evolutionary struggle, Ralph's tragic view resonates with how Michael L. Johnson describes the response to the end of the frontier that 'brought a recognition of disaster – and lamentation and nostalgia. Romantic excess led to tragic exegeses'.[36] Ralph projects, according to Johnson, a 'West of the mind', an image 'that had long been the fantasy of America's future'.[37] Indeed, Ralph and his class, so mired in nostalgia, appear to have *no* future.

Yet there was a view that regarded extinction as socially progressive. For example, as Bentley comments, Charlotte Perkins Gilman 'almost cheerfully embraces a motif of morbidity . . . [W]ith her faith in social evolution, Gilman foresees the birth of a new "economic free agent" out of the death of the "starved and thwarted class" of women, for the simple reason that "all morbid conditions tend to extinction"'.[38] As her pioneer blood befits her transformation into an 'economic free agent', Undine takes advantage of how post-frontier towns, such as Reno, Nevada, enable quick divorces that help women maintain their independence.[39] Even Ralph, moreover, wavers in his perspective on the effects of social evolution. On the one hand, the narrator, in a somewhat forced return to the water imagery of 'Undine' and 'Ararat', interjects that Ralph 'had never seen his way clearly in that dim underworld of affairs where men of the Moffatt and Driscoll type moved like shadowy destructive monsters beneath the darting small fry of the surface'.[40] From this cliché of social Darwinism, we move, a few pages later, to a moment when Ralph feels 'the thrill of the agent' over his investment in the Apex deal.[41] His more active posture implicates him in the ecological/economic transformation of the frontier, helping to bring about the changes that he so laments; a case, we might say, of 'imperial nostalgia'.[42] Yet this nostalgia is not mere fantasy, for it forecasts the environmental cost of the modernization of the post-frontier nation.

II. Business morality and Western water policy

Overall, the events in the novel surrounding the Pure Water Move seem to reference loosely the history of Western water management, in particular problems with contaminated water supplies. For example, the water in St Louis in 1893 'was often undrinkable'; 'the privately owned utilities were corrupt and inefficient'.[43] Even after improvements in time for the May 1904 World Fair, on the first day, as reported in the *St Louis Post Dispatch*, the lack of drinking water approached crisis proportions.[44]

In Chicago, at the end of the nineteenth century, the lack of potable water for a rapidly growing population caused 'frequent and virulent typhoid outbreaks'.[45] Chicago's solution – resonating with the legal–economic structures that Wharton depicts – was to stop using Lake Michigan 'as both a sewer and a water source' and build a channel (started in 1890) to transport waste through Lake Michigan to the Mississippi River, justifying this controversial decision to say the least (made after a one-year study by the Hering Commission) through the theory of 'self-purification': oxygen in the water 'purifies' waste and makes the water 'safe for consumption'.[46] Cities located in the path of the canal did not accept the validity of this theory; St Louis feared that any reputation for having polluted water would set it back in the economic competition in the region. A six-year battle in the Supreme Court over the canal, instigated by St Louis, resulted. The Sanitary District in Chicago funded scientists in order to build up evidence; and while making explicit its intentions, the organization did allow scientists freedom in carrying out research. In 1906, with the help of the strategic deployment of scientific evidence, Chicago won the case.

Replaying Wharton's script, in Kansas City the public water supply originated in 1873, with a franchise granted to the National Water Works Company of New York City. The Kansas River over time became extremely polluted, so the source was shifted to the Missouri River in Kansas, at a location 'approximately six miles' upriver.[47] The city bought the waterworks system in 1895. This purchase initiates a fascinating account of the development of the Kansas City water system, which warrants careful attention and emphasis.

The Western context for a philosophy of water management originated in the 'riparian doctrine' that protected landowners from any harmful influences upstream on the 'quantity' and 'quality' of their water.[48] In the late nineteenth century in Kansas, such an idea was complemented by the way that 'prior appropriation doctrine considered water as public property and recognized the right of a person to use some portion of a stream flow in a legally recognized "beneficial" use'.[49] This doctrine promoted seeing water as an economic resource.

There was, then, conflict during the Progressive Era over models for water management. For instance, debate arose over whether the Missouri River should be seen in terms of the criterion of 'economic' or 'healthy'.[50] The debate ranged along the lines of self-interest: the former view held by 'boosters of Missouri River navigation', the latter by those, at the 1910 Missouri River Sanitary Conference, who 'saw a hydrological system that linked people living upstream with those living downstream', as the river 'could either threaten public health or be used cooperatively as a public resource to the benefit of health'.[51]

The 1910 conference reflected the 'progressive political atmosphere' in Missouri and Kansas; the Kansas State Board of Health 'likely played the most significant role in instigating the conference'.[52] The Kansas Board in 1903 had set a national precedent by advocating 'state-level water pollution legislation', while Governor Herbert S.

Hadley of Missouri, though a booster of Kansas City business, was 'a progressive reformer who wished to control the power of corporate entities like Standard Oil and the railroads'.[53] Not only, however, was the conference a failure, due to a marked lack of attendance; its final report had little political influence.

Regardless, the conference marked an early attempt to address pollution in rivers. Major cities on the Missouri River pursued the cheapest method of 'filtering and treating drinking water at the waterworks' and dumped sewage, untreated, into the river.[54] Those who attended the conference were aware that outsized meatpacking companies in Omaha, St Joseph and Kansas City (Missouri and Kansas) dumped waste into the river, which contributed (albeit not directly) to the spread of typhoid; the rates of typhoid in Kansas City, Missouri, during the Progressive Era, 'were consistently among the worst in the nation'.[55] The conference reflected growing concerns about public health, in particular increased scepticism about water purification methods. Five years earlier, a Kansas City *Journal* editorial questioned 'the theory of dilution', promoted by a large number of engineers as 'an inexpensive and safe natural method of sewage disposal'.[56] Without precedent, the conference delegates proposed to have 'those states involved in the regional river system adopt common legislation', without federal oversight, 'to limit and prevent pollution'.[57]

In 1915, the Missouri Valley Public Health Association (MVPHA) held a conference that addressed pollution in the river. The delegates shifted responsibility from major cities to farms and small towns to maintain pure drinking water. Rather than the 'volume and content of urban and industrial wastes disposed of in the river', the 'potential disease-carrying' flies found on farmers' refuse in the river was (remarkably) claimed to be the greatest public health threat.[58] The MVPHA president, Dr Paul Paquin, the Kansas City public health director, favoured a policy (in 1915) that emphasized individual incentive to kill flies that could be a vector for diseases such as typhoid. Such a policy was indeed 'curious', as 'larger municipal programs were likely to result in a better public health record'.[59] Focusing on specific issues of water pollution (a less holistic vision), the MVPHA meshed models of a healthy river and an economic river (having in common a desire to look past 'political boundaries' by connecting communities upstream and downstream), though the models had different priorities and areas of focus: the former sought reform through 'interstate agreements', the latter 'at the federal level'.[60] Importantly,

> the cost of economic development had casualties. Those who lived in urban river districts – like immigrants, minorities and the working poor – would not participate equally in the wealth gained from river transport. These were the same populations most stricken by the effects of river pollution because they did not have access to good drinking water and other municipal services.[61]

As a brief historical survey suggests, the opposition between biopower and business morality would not be mediated easily or rapidly.

III. 'Constitutional restlessness' and 'something not ourselves'

If Wharton's Pure Water Move is a satirical commentary on how the moral attitudes that shape environmental policies are culturally 'contaminated', then we will now observe how her commentary moves from addressing a specific case of environmental biopolitics to a larger investigation of ethics in the modern era. Through the episteme of modernity, ethics become redefined through a regime of emerging, autopoietic (self-making) complex systems – for instance, Wall Street epitomizes the interpenetration of economic, legal and information systems – that has important effects on the idea of performing identities at the individual and national level. In his reading of systems theorist Niklas Luhmann, Wolfe concentrates on the idea of functional differentiation: to deal with environmental complexity, autopoietic systems differentiate themselves into subsystems, a process that reinforces the system/environment distinction with systems constituted as environments for their subsystems. In Wharton's novel, class differences produce stratified social subsystems: what the old money disdained 'were the centre of social systems far outside its ken, and as indifferent to its opinions as the constellations to the reckonings of the astronomers'.[62] The functional differentiation of social systems, according to Wolfe, resists 'some untainted, unmediated space of subjunctive interiority . . . some unquestioned and unquestionable yardstick by which the socially and linguistically constructed code of ethics may be judged'.[63] That is, because ethics can only be observed through a system that self-maintains through its own operating codes, 'the distinction between morality and merely ethical "roles" is lost'.[64] Ethics come to be regarded as performative, defined through a constant state of becoming.

While collectively ideas about performance and ethics have been rehearsed in the previous chapters, an early image of Undine in *The*

Custom of the Country suggests a further transformation of these ideas: 'She was always doubling and twisting on herself'.[65] Again, according to Judith Butler, '[P]erformativity is thus not a singular "act", for it is always a reiteration of a norm or set of norms', while at the same time, '[R]eiterations are never simply replicas of the same'.[66] Undine's performance is based on 'reiteration', having 'an instinct of adapting herself to whatever company she was in, of copying "the others" in speech and gesture as closely as she reflected them in dress'.[67] Another example from the novel points to Undine's performance as destabilizing, even as she 'becomes' her 'self', putting the idea of her self-identity, who she is, into question: 'Undine was fiercely independent and yet passionately imitative. She wanted to surprise everyone by her dash and originality, but could not help modeling herself on the last person she met'.[68] Her notion of desire as a signifier of self-will is also reiterative; as she states, 'I want what the others want'.[69] Rather than exhibiting an authentic self, she suggests that the idea of authenticity is created through a social performance that is guided by what others want and how others want to see her. Undine's performance resonates with Butler's comment that 'the impossibility of a full recognition, that is, of ever fully inhabiting the name by which one's social identity is inaugurated and mobilized, implies the instability and incompleteness of subject-formation'.[70]

Social mobility in the novel, especially the free agency of women, is sustained by the viewing of marriage and divorce as economic transactions. Looking at the new social class represented by Undine, Ralph muses, the female members 'bought their husbands as they bought an operabox. It ought to have been transacted on the Stock Exchange'.[71] Undine's 'inalienable right to "go around"',[72] her agency, achieved through social and economic modernization, is however created by her being bound to social conventions rather than transcending them. '[I]f there is agency', Butler thinks, 'it is to be found, paradoxically, in the possibilities opened up in and by that constrained appropriation of the regulatory law, the compulsory appropriation and identification with those normative demands'.[73] Even as she masters what Bentley calls 'the American system of "easy divorces"',[74] Undine's free agent status underlines how divorce is constructed through the social contract of marriage. In the novel, Charles Bowen states, 'one couldn't be divorced without it'.[75] Defining marriage as a performative speech act, Butler explains: 'The centrality of the marriage ceremony in J. L. Austin's examples of performativity suggest that the heterosexualization of the social bond is the paradigmatic form for those speech acts which bring about what they name. "I pronounce you..." puts into effect the relation it names'.[76] Thus Undine, as a divorcée, is still 'named' by the 'social bond' of marriage.

In the early twentieth century, the meaning of Undine's performance, however, was viewed with concern. As Bentley chronicles, anthropological studies and museum exhibits reflect an anxiety about female social agency, represented as 'the atavistic power of women to disrupt – even to reverse – the advances of society'.[77] Bentley's interpretation of the novel demonstrates how the upper class manages this anxiety through the privileging of a detached, scientific view that constructs a 'realistic' image of women. Supposedly produced from impartial observation, this image characterizes many photographs and portraits of the period. Like her use of marriage, Undine advances socially through her submission to the male gaze, acting as its object of desire by participating in social spectacles – posing for photographs, being featured in newspaper columns consumed by a mass audience – that expose Undine to greater visibility, thereby increasing her influence. Bentley posits that 'the erotics of detachment in realism, by denying its full interest in mimetic spectacles, makes the realist vision vulnerable to being taken in by its own desire – a desire that prompts the detachment in the first place'.[78] Undine's appropriation of this desire troubles the supposed detachment of observation, questioning in turn the acts of scientific observation that social reporting, striving to be credited as objective truth, imitates.

Hence Wharton's portrayal of natural selection satirizes the belief that evolution acts on a helpless and passive 'victim', who loses out in the struggle for survival. For example, Bowen's observation of what he thinks is the evolutionary failure of Ralph, set against a nouveau riche backdrop of 'plumed and jewelled heads', expresses 'the pang of the sociologist over the individual havoc wrought by every social readjustment: it had so long been clear to him that poor Ralph was a survival [i.e. relic], and destined, as such, to go down in any conflict with the rising forces'.[79] The heading of the stationery on which he writes his thoughts, 'The Parisian Diamond Company – Anglo-American Branch',[80] however, questions his apparent detachment. Bentley remarks: 'Bowen is an agent for the economic circuit that joins American capital with Parisian culture – and, in all likelihood, with the raw materials and labor of colonial Africa – to produce the spectacular social order he views with such wry detachment. The "jewelled heads" under his gaze are incarnations of the diamond capital he serves'.[81] Furthermore, such capital was on the rise, as diamonds became a signifier of marriage in the early twentieth century.[82] Wharton's critical evaluation of natural selection undermines the validity of objective reporting, as everyone in the novel is implicated in the social and economic changes that inform Undine's ascent. Unlike, for instance, Gilman, who viewed extinction as a 'natural' outcome of social progress, Wharton illustrates how the idea of evolutionary decline is validated through cultural beliefs, that is, upper-class anxieties about the end of its social reign. Indeed, Ralph's evolutionary fantasy, discussed earlier, can be interpreted as a satire of determinism, for he adheres to the idea of a biologically essentialist identity, in contrast to Undine's 'fitness' signified by her ability to perform a dynamic, unbounded self. To put it somewhat differently, he is an unskilled performer. Daunted by Moffatt's staging of the Apex land transaction, Ralph recalls when 'he once attended a lesson in acting given at the Conservatoire by one of the great lights of the theatre'; he now 'began to feel as blank and blundering as the class of dramatic students before whom the great actor had analyzed his part'.[83]

Being a free agent informs Undine's performance, in images such as her '"play[ing] lady" before the wardrobe mirror' and rehearsing for her moment on the social stage: 'Within a few days she would be enacting the scene she was now mimicking'.[84] Yet Undine's performance as a free agent has national dimensions. The French Duchess remarks of Undine, '[S]he's American, she's divorced', stated 'as if she were merely stating the same fact in two different ways'.[85] At the same time, Undine's self-deconstruction pressures a stable national identity. For Bentley posits, there is no 'real' Undine; her performance becomes 'uncanny', enacting her *alien-nation*.[86]

Priscilla Wald takes Freud's study of the uncanny (from *unheimlich*, 'not homelike') to be central to 'official stories of the nation'.[87] As discussed in Chapter 3, she elaborates that the concept 'turns on the discovery that the unfamiliar is really familiar (the stranger as self) but also that the familiar is unfamiliar (the self as stranger) . . . [T]he uncanny sends us home to the discovery that "home" is not what or where we think it is and that we, by extension, are not who or what we think we are'.[88] Part of an American group of 'westward bound nomads', Undine appears adrift in Europe.[89] Yet even at 'home' in the US, a destabilized national identity persists, intensified by the post-frontier West – as Raymond has earlier noticed. Bowen's observations, while they disclose upper-class concerns about social authenticity, frame Wharton's depiction of the American divorcée and the spectre-like global national economy as a product of the uncanny: 'The dining-room at the Nouveau Luxe represented, on such a spring evening, what unbounded material power had devised for the delusion of its leisure: a phantom "society", with all the rules, smirks, gestures of its model, but evoked out of promiscuity and incoherence while the other had been the product of continuity and choice'.[90]

What Butler describes as the 'instability and incompleteness of subject formation' connects to the emerging performance of the US on the global stage. The novel, composed during 'the take-off phase of modern globalization', reflects desires for national expansion and concerns about the instability of the national subject.[91] Wharton shared these views with Archibald Cary Coolidge, a Harvard historian whom she met through her husband, and William Morton Fullerton, a political journalist with whom she had an affair. Coolidge's *The United States as a World Power* (1909) asserts that American subject formation is unresolved: '[T]he nation is still "in the making"'.[92] To Coolidge, anxieties that the US might devolve into a 'country inhabited by many different and often discordant nationalities' can be treated by assimilating immigrants into the nation, which maintains an inner stability while extending further its boundaries.[93] Fullerton's *Problems of Power* (1913) explains how US expansion that fuels feelings of national disorder contributes to desires for nationalism: '*national spirit is manifested only when nationality is menaced*'.[94] For Fullerton, the source of the conflict is

> the predominance to-day of economic over political conditions. The problem of the maintenance of national traditions, national characteristics, national integrity implies a constant compromise between world-interests, human interests in general, on the one hand, irrespective of national classifications, and, on the other hand, those sentimental, hereditary, beautifully persistent impulses and prejudices that shape a nation's soul.[95]

Both Coolidge and Fullerton try to balance encouraging US imperialism as a continuation of Manifest Destiny and maintaining national stability.[96] But Wharton's novel seems more critical of this negotiation, demonstrating that, through the uncanny, we find merely a ghostly signifier of national fantasies of global leadership.

Undine's performance, moreover, references the nation's founding documents: her argument for her social independence that echoes 'school-pieces about freedom and the British tyrant', her investment in an 'inalienable right to "go around"' and her 'constitutional restlessness'.[97] As Undine restages the constant state of 'becoming' a national subject, she comments on the instability of being American that is, always and already, on the scene of national development. Barry Shank's analysis of the Constitution, a 'naming' document for the nation, critically examines the 'three-fifths' compromise: '[F]or the purpose of taxation and representation . . . non-free persons – slaves, the legal opposite of independent economic agents – had to be counted. And in that counting, they were imagined as not citizens . . . but as fractions of persons'.[98] Because race is not named in the critical phrase from Article I ('three fifths of all other Persons'), Shank insists that 'this very phrase, in its abstract blankness, establishes the profound performativity of the nation conjured into being by the document, requiring that the question of citizenship be continually reopened and reperformed, legally, procedurally, and culturally'. That the issue of race haunts the 'constitution' of American identity is suggested repeatedly in Wharton's novel. Undine's desire to be named as a free agent takes advantage of opportunities that are restricted by race. Further, the US as participant in the global marketplace is implicated in racial exploitation, as the heading on Bowen's stationery ('The Parisian Diamond Company – Anglo-American Branch') reminds us. The US economy, in the past rooted in slavery, now exerts a global influence. The economic expansion that helps to underwrite Undine's social freedom as an American divorcée, at the same time, threatens to exploit those overseas who remain colonial subjects.

Unresolved questions about race and citizenship compel the reperforming of American identity. There is, according to Shank, 'the endlessly repetitive need to reiterate what it means to be American'. Such a need very much concerns Coolidge. While for Frederick Wegener it represents 'a distinctly American exercise in colonialist apologetics',[99] Coolidge's *The United States as a World Power* does to some degree suggest the mounting costs of imperialism, noting that racial issues persist both outside and inside the nation's borders. As for having to deal with racial tensions, Coolidge warns, 'every nation holding colonies will have to face such problems sooner or later'.[100] An even more pressing concern is how to deal with the legacy of slavery: 'Of these populations, only an insignificant fraction represents the original dispossessed inhabitants: the vast majority have inherited an even worse grievance, for they are the descendants of imported slaves. The proper treatment of these peoples is a matter of momentous importance for the future of the republic'.[101] Granted, Coolidge's response to questions of race is at best evasive, but at the very least he does emphasize the importance of answering such questions: nothing less than 'the future of the republic' is at stake.

The ongoing inability to reduce tensions between race and citizenship becomes especially visible in legal discourse. As Wald elaborates, the Supreme Court decisions on *Cherokee Nation v. the State of Georgia* (1831) and *Scott v. Sanford, or Dred Scott* (1857) left unresolved the identity of indigenous Native Americans and African descendants, wavering between the legal status of 'citizens' and 'aliens'.[102] A similar verdict, at the turn of the twentieth century, was announced concerning the status of Puerto Ricans in *Downes v. Bidwell* (1901). To Wald, these cases disclose that 'the emerging imperial nation once again legislated its own ghosts into existence in the name of the (white) American family, and once again these presences were haunted by virtue of their resemblances as forcefully as their differences'.[103] Brought about through the desires of the 'imperial nation', Wharton's 'phantom' society, as a product of national global expansion, reflects the indeterminate legal logic that continually leaves open questions about the meaning of American citizenship.

The collective voice of 'We the People', rather than self-possession, reveals the 'possession' or 'haunting' by past and present ghosts. Shank goes on to propose that 'We' should be read in the same way as 'I' in Butler's argument, where, in her 'paradoxical' account of subject making, the subject has to be called into being or named from beyond and before any act of self-expression: 'I can only say "I" to the extent that I have been first addressed, and that address has mobilized my place in speech'.[104] Accordingly, as Shank reads Butler, we see how the performativity of language displaces the national subject. In Butler's words,

> If a performative provisionally succeeds (and I will suggest that 'success' is always and only provisional), then it is not because an intention successfully governs the action of speech, but only because that action echoes prior actions, and *accumulates the force of authority through the repetition or citation of a prior, authoritative set of practices*. What this means, then, is that a performative 'works' to the extent that it *draws on and covers over* the constitutive conventions by which it

> is mobilized. In this sense, no term or statement can function performatively without the accumulating and dissimulating historicity of force.[105]

The absence of race in naming the national 'We' causes the instability of being American and can only be 'covered over' by the 'conventions' that justify a partial naming; that is, through a restrictive and arbitrary process, some national subjects are named as citizens. In *The Custom of the Country*, Undine's performance is not the only marker of self-estrangement. Ralph expresses similar feelings of alienation when he compares his family to Aborigines, who have been displaced by economic and social change. While Ralph's 'playing Indian' counterparts Undine's 'pioneer blood', it restages, at the same time, the question of Native American citizenship, invoking another racialized ghost of the past.[106] But Ralph is not the only one who plays Indian to express concerns about the namelessness of American identity. Harvard English professor Barrett Wendell, at the turn of the twentieth century, writes: 'The very name of us means something not ourselves . . . I feel as I should think an Indian might feel, in the face of ourselves that were'.[107] Wendell's response echoes the exaggerated feelings of loss that Ralph feels. As Wald comments: 'In a disturbing analogy, Wendell summons the destruction of not just of another culture, but of the very culture ("American") responsible for the original destruction'.[108]

The imperial desires of the US that shaped new forms of international relationships increase anxieties about national subjectivity. In the novel, the marriage between Hubert de Chelles and the American Looty Arlington is referred to by the American newspapers as 'another brilliant foreign alliance'.[109] But what kind of American would these foreign alliances produce? This question is underlined by Theodore Roosevelt: '[T]he man who does not become Americanized nevertheless fails to remain a European and becomes nothing at all'.[110] Considered together, Wendell and Roosevelt articulate the uneasy feelings at the turn of the twentieth century about the making of what Laurent Berlant calls 'national fantasy', that is, 'how national cultures becomes local – through the images, narratives, monuments, and sites that circulate through personal/collective consciousness'.[111] Because imperialist pressures disrupt the production of national fantasy around 'We the People', it means that we can no longer say who we *really* are, because the name 'American' seems distanced from *being* American: we are 'something not ourselves'. To put it in terms of Butler's argument, such pressures make us aware that the meaning of 'American', which can create an authentic experience of national citizenship, never arrives, for that performative name depends on a prior citationality that both invokes and conceals the unresolved racial traumas of national history.

But for Karen Barad, Butler's argument does not go far enough. Barad contends:

> Butler's concern is limited to the production of human bodies (and only certain aspects of their production, at that), and her theorization of materialization is parasitic on Foucault's notions of regulatory power and discursive practices, which are limited to the domain of human social practices. Furthermore, for both Butler and Foucault, agency belongs only to the human domain, and neither addresses the nature of technoscientific practices and their profoundly productive effects on human bodies, as well as the ways in which these practices are deeply implicated in what constitutes the human, and more generally the workings of power.[112]

The fantasy around the constitution of the national subject hides another fantasy: that somehow 'the production of human bodies' stands apart from the natural realm. In Wharton's novel, such an invented distance justifies looking at natural resources as sites for the exercise of national power – which extends to the global stage, as the heading on Bowen's stationery suggests. The Pure Water Move, because(as Seltzer reminds us) it exerts control over natural resources, is just the sort of cumulative example of 'technoscientific practices' that Butler and Foucault overlook. In its current form of bioethics, biopower, as Wolfe reminds us, tends to be suspect, for it is rooted in the disciplinary workings of 'regulatory power and discursive practices'.[113] As a result, the standing of 'moral attitudes' appears in some way separate from 'social norms'. What Barad proposes is to move away from this sort of fantasy, as a discursive construction, and towards a rethinking of matter itself as active instead of passive. Opposing (national) fantasy to what she calls an 'agential realist account', she writes that 'matter does not refer to a fixed substance': it is '*not a thing but a doing, a congealing of agency*'.[114] To put it broadly, in the context of Wharton's narrative, Undine's performance does not just destabilize identity as an individual and national formation; it provides a critique of the realist project, in how that project renders the world *visible*, which, in its *performativity*, has implications for theorizing biopower. I will now take up this critique.

IV. Systems of art: perception and communication

From a systems-theoretical perspective, art expresses such a critique in a unique way, as a question of observation and then meaning. Of Popple, the society artist, whose upper-class patronage depends on his ability to 'do pearls', we hear Ralph state: '[W]hen he began a portrait he always turned back his cuffs and said: "Ladies and gentlemen, you can see there's absolutely nothing here"'.[115] Of the shift away from emphasis on the content of art, Bentley explains,

> [A] Popple portrait is not meant to represent a person – an individual's face, hair, clothing, a personal aura – so much as to display social prestige, and pearls are therefore one of a female portrait subject's most distinguishing features. To do pearls is not just to paint them, it is also to use an aesthetic image to convert an economic object into a conspicuous signature of status.[116]

There is a different way, however, to describe this shift. According to Wolfe's reading of Luhmann, art as a social system responds to modernity (epitomized by Undine as free agent), a condition that creates the functional differentiation of systems. Wolfe explains:

> [A]ll systems carry out their operations and maintain their autopoiesis by deploying a constitutive distinction, and a code based on it, that in principle could be otherwise. This means that there is a paradoxical identity between the two sides of the system's constitutive distinction, because the distinction between both sides is a product of only one side. In the legal system, for example, the distinction between the two sides legal/illegal is instantiated . . . on only one side of the distinction, namely, the legal. But no system can acknowledge this paradoxical identity of difference – which is also in another sense simply the contingency – of its own constitutive distinction *and* at the same time use that distinction to carry out its operations. It must remain 'blind' to the very distinction that links it to its environment.[117]

We are on the way to seeing the form of art as not content, but as observation – especially observing observation (second-order observation) that allows us to decode this 'blindness', or what Popple calls 'nothing', and recognizes, without our directly seeing it, our own space of invisibility (what we cannot see). For Luhmann, art 'imitates the world's invisibility, a nature that can no longer be apprehended as a whole'.[118]

This theme surfaces in the Italian honeymoon vacation of Ralph and Undine. Financially pressured, due to the rushed wedding, their relationship is beginning to wane. Ralph, trying to balance and satisfy his desires for 'money' and 'harmony', has a vision, while watching the moon rise one night, of a book he could write.[119] This is the resulting exchange between Undine and Ralph:

> 'A vision?' Her interest flagged. 'I never cared much about spirits. Mother used to try to drag me to séances – but they always made me sleepy.'
>
> Ralph laughed. 'I don't mean a dead spirit but a living one!'[120]

Given that Ralph has already been, and Undine will be, associated with 'dead spirits' (in particular in the portrait scene in Popple's studio), Ralph's artistic inspiration seems rather thin. The almost clichéd response Ralph has to the pastoral landscape of Italy appears not as an alternative to Popple's painting, but rather a global extension of it, as Ralph's supposedly authentic creative impulse is soon reduced to developing 'what the masterpieces of literature had mostly been – a pot-boiler'.[121] The economic negotiations that inform artistic and literary production appear further amplified when Ralph, later feeling the 'thrill of the agent', thinks of Moffatt as a 'novelist'.[122] Then, at his nadir (after the financially and emotionally draining divorce proceedings), Ralph revises his literary vision: 'He no longer saw life on the heroic scale: he wanted to do something in which men should look no bigger than the insects they were'.[123] While Ralph, in effect, seeks to imitate the naturalist register of Wharton's narrative, he adheres to the social Darwinist doctrine that Wharton's social satire so critically interrogates. Hence the blind spot of an observation (as we have been discussing) renders contingent the ways in which art can be said to *imitate* nature.[124]

That art brings with it an intrinsic negation of vision (imitates the world's invisibility) suggests that, although Popple may be a hack artist, he intuits that there is something beyond the formal aesthetic of his painting. How he describes his artistic process (as 'absolutely nothing') would explain why his studio is not set up as an 'art space' per se – a place for contemplation, working and production – but rather as an ongoing tea or cocktail party complete with a bar, a place for (fleeting) forms of bourgeois pleasure.[125] This moreover puts observers like Bowen in a tight spot, signalled by his 'amusement of watching the picture compose itself *again* before his eyes':[126] the 'real' society (as a supposed product of 'continuity'), as it were, now seems to differ little from the 'phantom' society, a difference that so matters to the realist project. This realization, as indicated by the Italy scene, goes beyond the confines of Popple's studio, for Wolfe comments that 'all observations, including those of nature, are contingent *and* of necessity blind to their own contingency'.[127] That said, Popple's studio is the focal point of our discussion, as I have been arguing, because Popple puts the emphasis on what he does (doing pearls, for instance), the art-making as becoming (which displaces vision, creating 'nothing'), rather than the formal content of the art itself. As Luhmann would put it, it is a case of 'presenting the art of presentation itself', which 'amounts to creating specific forms for an observation of observations'.[128] In this light, Peter and Undine trace the observing of the economic transactions beyond the portrait which shape its creation. Peter exclaims that 'the pearls ain't big enough'; Undine replies that Popple is not to blame, because Ralph, *not* Popple, gave the pearls to her, suggesting the 'real' pearls are unsatisfactory, before speculating on what the portrait of her will 'cost'. Popple interjects that the question of observing what is 'real' is abstract: '[T]he artist, you understand, sees all this as a pure question of colour, of pattern'.[129]

By thinking (even briefly) about the different viewpoints for how to regard art, those of Peter, Undine and Popple, we can start to appreciate the difference between perception and communication. Wolfe elaborates:

> [A]lthough both perception and communication operate as autopoietic systems on the basis of difference and distinction, the very different speeds of processing of these systems makes it appear that perception confirms, stabilizes, and makes immediate, while communication (to put it in Derridean terms) differs, defers, temporalizes. In the work of art, the *difference* between perception and communication is re-entered in the services of the work's communication. But because of this asymmetry in speeds, it is re-entered in a way that calls attention to the contingency of communication – not of the first-order communication of the artwork, which appears incapable of

> emendation (it is what it is), but of the second-order observation of the work's meaning and its critical function vis-à-vis the system of art.[130]

Works of art have 'something like a privileged critical relationship to society', according to Wolfe, because 'they literally fix social space'.[131] In the novel, Popple's studio is 'privileged' because of the art therein that 'fixes' social space as a stage for performance; the portrait itself becomes a mediated performance. Popple's studio, as will be made clear, connects the diverse social spaces in the novel (the Pure Water Move and Apex deal, the property/propriety of old money and new, the Ararat investigation); as Luhmann says of works of art, 'their significance as objects implies a realm of social regulation', a realm that emerges in particular through 'the more complex case where one observes other observers by focusing on the same object'.[132]

The viewing of Undine's portrait, attended by her social circle, continues and sharpens the conversation about art and realism. With his irreverent attitude on full display, Peter exclaims,

> Those big chaps who blow about what they call realism – how do *their* portraits look in a drawing-room? Do you suppose they ever ask themselves that? *They* don't care – they're not going to live with the things! And what do they know of drawing-rooms, anyhow? Lots of them haven't even got a dress suit.[133]

Peter's words indicate the social regulatory at work, as taste and custom influence the work's meaning and its critical function. Popple's response seems to insist that the meaning of art is first- rather than second-order observation: '[T]here has been no need to idealize – nature herself has outdone the artist's dream'.[134] In the novel's logic, Popple's portrait of Undine, which as Bentley notes 'displays social prestige', rather than the self, is expressing the social regulatory as its meaning, which is of course 'natural', yet not in the way Popple thinks or intends. Because, as we now know, nature is contingent; it 'can no longer be apprehended as a whole'.

Such knowledge circles back to Undine's self-performance – as an act of becoming, it is contingent – and especially so when it reveals the instability of the social order. There is the association between Undine's casting herself in the portrait 'in a "dead white" Pre-Raphaelite glamour' and what Moffatt tells Abner, that the acquisition (by Driscoll) of the Apex water resources is 'a dead sure thing'.[135]

FIGURE 5 John William Waterhouse, 'Undine'

The association is extended further when Undine disseminates 'the success of the picture' by planning to 'telephone her press-agent to do a paragraph about Popple's tea',[136] suggesting how the mass media convey a sense of reality defined through taste and custom – an attempt, we might say, to mitigate anxieties about what is 'sure' in a contingent world. Yet this moment of social perception, so to speak, can only be constituted through art doing what it does, art as the communication of Popple's earlier 'pure question', stated otherwise as the question of what is pure. A few pages later, talk about Driscoll's problems (his father, as his source of wealth, is a target of the Ararat investigation) allows Peter to further his efforts to seduce Undine (Peter offers to put on the ball himself). Undine's response to Peter's bid of seduction re-presents, through performance, the social regulatory: '[H]er impatience to enjoy was curbed by an instinct for holding off and biding her time that resembled the patient skill with which her father had conducted the sale of his "bad" real estate in the Pure Water Move days'.[137] As an image of biopower, the Pure Water Move reminds us how business morality (selling '"bad" real estate') is connected to social/sexual scandal. The Pure Water Move is framed by Moffatt's elopement with Undine, and re-presented now as a frame for Peter's potential affair with Undine. Undine continues the negotiations, if you will, with Peter that are often explicitly tied to money, because 'it was amusing, too, to be "talked about" with Peter Van Degen'.[138] Undine, her agency staged in Popple's studio, is able to use social/sexual scandal for her own ends, a strategy that (re)maps the relays of (sexual) pleasure, business morality and social regulation.

Yet business morality attempts to distance itself from any hints of sexual indiscretion. In the words of Bowen: 'In America the real *crime passionnel* is a "big steal" – there's more excitement in wrecking railways than homes'.[139] Accordingly, Abner is stunned when he finds out that Peter has given Undine a set of pearls (bigger than those in the portrait) after their liaison, and insists that Undine return them. But Undine thinks differently. According to Bentley: 'She refuses the idea that the pearls are a payment for sexual services and treats them as an investment in a new marriage'.[140] What results is a marriage not to Peter, but to Raymond de Chelles. Thus Undine's performance, befitting her name, of a 'fluid' identity and her agency, seen as social 'circulation', transforms the impure (sex for money) into the pure (the ideal of marriage). Similarly, the Pure Water Move, as constituted through (in Seltzer's terms) the technoscientific 'channelling' of force (through biopower), changes bad land into a benefit for the public good, while the theories of 'self-purification' (Chicago water policy) and 'dilution' (Kansas City policy) transmute contaminated water into potable water. Indeed, the pure question, expressed through Undine's portrait, becomes an 'unavoidable provocation to *another* communication whose meaning is far from immediately clear', as Wolfe would say, that is, the question of the social regulatory at work.[141] *That* question is staged as the way the image of the pearls (as biological wealth – the only gem produced from living aquatic creatures) in the portrait ultimately discloses that the tighter the hold on custom and taste that the social regulatory has, the more its *own* contingency (as a product, Luhmann reminds us, of second-order observation) comes into view.

V. Pure fiction

That is not to say, however, that to call something contingent rules out what Wolfe would call 'its material effects on other living beings'.[142] As in *The House of Mirth,* the materiality of text constitutes an official account: the newspaper clippings of Undine's divorce from de Chelles and (re)marriage to Moffatt. While these clippings, as Bentley points out, certainly articulate Undine's social influence, they also reprise an ongoing biopolitical narrative: the cultural struggle to exert (modern) control over sex and embodied lives through marriage. In this case, Moffatt's gift of a 'pearl necklace' to Undine discloses Moffatt's sway over the 'price of pearls'.[143] At the end of the novel, biopower resonates at a global level: during the brief marriage, de Chelles has had to 'rebuild' a dam in France that protected his agricultural holdings.[144] Major players from the Pure Water Move return; the news that Jim Driscoll's wife has been made an ambassadress (perhaps the result of another backroom deal) reminds us of Driscoll's political survival instincts that leverage concerns about public health. What Bentley, moreover, calls Undine's 'monstrous' agency stages a central conflict in biopolitical thinking: between what Eugene Thacker describes as Foucault's objective 'to allow circulations and flows' and a reading of biopower that shifts into its deathly function – that is, the state regulation of death.[145] The latter, alluded to by Hurstwood's suicide in *Sister Carrie*, emerges in *The Custom of the Country* through the images of 'still life' that connect Undine's deathly portrait and Moffatt's reference to the acquisition of the Apex water resources as a 'dead sure thing'.

While Undine, as an economic free agent, appears to rewrite the business morality of her father, which governs financial speculation and romantic relationships (albeit unequally), the realm of social regulation returns with a vengeance at the end of the novel: an oppressive marriage to de Chelles, a hasty remarriage to Moffatt (whom she ultimately finds tiresome and unattractive), the prohibition against her (as a former divorcée) becoming an ambassadress – 'the one part she was really made for'.[146] As to the final prohibition, we may feel relieved that the contingency of the social regulatory limits the realm of speculation (Undine cannot financially negotiate becoming an ambassadress), but such feeling comes with the cost that social/environmental regulation as biopower, as a *sure* distinction between pure and impure, can have no ethical grounding. And because the question of ethics is open, the nightmare of ecological breakdown must remain with us: Luhmann bluntly tells us that 'systems pursuing a trend toward exposure to ecological self-endangerment are eliminated'.[147] Put in the context of Wharton's novel that flirts with apocalypse (the scandalized Ararat Trust signifying a natural/national landscape 'after the flood'), the 'thrill of the agent' is limited due to the 'material effects' of *pure fiction*.[148]

5

THE MISEDUCATION OF HENRY ADAMS: FANTASIES OF RACE, CITIZENSHIP AND BIOLOGICAL DYNAMOS

One aspect of popular scientific journalism that we have been exploring comes to the fore regarding thermodynamics, and the physical sciences at large: a rhetorical strategy that justifies scientific theory and practice by lining them up with daily lived experiences. W. S. Franklin's 'The Second Law of Thermodynamics: its basis in intuition and common sense', published in *Popular Science Monthly* in March 1910, is the forerunner of a more specialized article on entropy by the same author three months later (in *The Physical Review*, June 1910), its approach from the start detailing the widespread importance of thermodynamic concepts that exceeds disciplinary boundaries. Franklin writes that 'no generalization of modern physics is based upon such deeply seated and such widely diffused human institutions'.[1] He places thermodynamics within the larger schema of the physical sciences, where the aim is to 'develop conceptions of the physical processes of nature' and 'correlate phenomena on the basis of sensible things, things that can be seen and measured'.[2] From here, however, Franklin deals with abstract ideas such as entropy by reclassifying them: this science 'refers to the things of the kitchen and to the things of the furnace . . . which come out of those mysterious places where maids and furnace-men rule'.[3] Franklin draws from the nomenclature of working-class labour practices to underline the pragmatic objectives of the scientific approach, for 'the most practical thing in connection with the turbulent aspect of any physical phenomenon is its general result or consequence'.[4]

Franklin is not, I would point out, being patronizing towards class interests here. Rather, he takes seriously the materialistic aspects of science: thermodynamic processes do indeed act on and through human bodies, theoretical notions of work and energy having ready analogues in daily lived experiences. Yet these experiences are circumscribed by capitalist ideology, a point that Franklin reminds us of in his account of how economics shapes the popular understanding of thermodynamics:

> In every-day life we see the fire-insurance companies concerned with certain broad features of statistical physics in their examinations and records of fires, and we see them also concerned with a profit and loss account which is wholly abstracted from the details of the phenomena of conflagration. Thermodynamics is the profit and loss branch of physics as it were; and like the profit and loss branch of fire insurance, thermodynamics is completely abstracted from any consideration of the details of any physical phenomenon.[5]

Franklin's emphasis on 'loss', 'degeneration' and 'waste' reads as the opposite of Herbert Spencer's evolutionary optimism.[6] Furthermore, fire-insurance calculations recall, to a degree, Wall Street speculation, responsible in part for the economic failure in the late 1890s, a feature of the recent past of which readers would have been aware. Thermodynamics, in Franklin's framing, becomes a science (the law not of conservation but of waste) well suited to complement and explain the economic disruptions that would be inherent to a rapidly expanding global imperialism, 'the records of fires' also historical markers of conquest and signifiers of the creative destruction that drives economic speculation.

In *The Education of Henry Adams* (1907), Henry Adams describes his experience of financial crisis, which arguably shapes his distrust of scenarios of evolutionary progress as much as his finding scientific and political evidence that could challenge Spencerian optimism. Certainly, Adams is prone to embrace thermodynamics, because it fulfils his view of modernity as a historical period of degeneration and collapse. But Franklin also outlines the larger objective that guides Adams's scientific speculations: an attempt to generalize history into systems that operate according to scientific principles. What Adams is doing in his use of ideas such as entropy and phase-change, then, is fashioning a lens for observation that could capture the complexity, chaos and dynamism of his social environment.

Monkey-wrenching Adams's scheme, Thomas Pynchon's novel *Against the Day* (2006) opens with a visit by the boy aeronaut adventurers, The Chums of Chance, to the 1893 Chicago Exposition. This episode seems designed to reference one of the central events in Adams's *Education*, the author's scientific awakening at the Exposition on witnessing the power of the dynamo.[7] Yet given Pynchon's sly mimicry – the Chums recalling Adams's 'boyish' wonder about science, their policy of non-involvement with the civilizations they observe echoing Adams's self-proclaimed stance of worldly detachment – the dynamo is curiously absent. Instead, Pynchon focuses on their aerial view of the Chicago stockyards:

> [I]t was as if the Chums, who out on adventures past, had often witnessed the vast herds of cattle adrift in everchanging cloudlike patterns across the Western plains, here saw that unshaped freedom being rationalized into movement only in straight lines and at right angles and a progressive reduction of choices, until the final turn through the final gate that led to the killing floor.[8]

Pynchon's response critiques Adams's rush to applaud the exotic and self-ordering dynamo, for Adams overlooks the more mundane but more numerous instruments for maintaining order. Furthermore, the 'freedom' of cattle turned into captivity alludes to the loss of the open frontier effected by Manifest Destiny. Thus the opening of *Against the Day* brings out in the open Adams's tendency to repress his fears

FIGURE 6 Chicago stockyards, 1909

about control. To some degree, Adams sublimates his social concerns about 'a progressive reduction of choices' through his scientific obsession with the loss of usable energy in a closed system, that is, the Second Law of Thermodynamics. In an entropic state, conditions of absolute order and chaos could be said to be equivalent; that this fact eludes Adams helps to illustrate the noticeable gaps in his scientific education, which he tries to fill using makeshift models, such as his 'Dynamic Theory of History'.

Now, an obsession with deterministic ideas is a standard feature of US naturalist writers. What makes Adams different, however, is that his concerns shift from mechanical to informational control. And this shift occurs during a crucial change in US foreign policy as an aggressive stance of taking overseas regions by force is muted into a subtler form of imperial coercion, maintained by informational systems that range from the transatlantic to the global. In what follows, I will show how Adams's *Education* uses racial, gender, class and national tropes to make connections among evolution, thermodynamics and informatics. Adams pushes against the standard evolutionary thinking of his time by introducing a thermodynamic framework, pressuring the doctrines both of Charles Darwin's gradualism and Spencer's teleology. The question that Adams repeatedly asks – who is, can be and will be American – must be evaluated against the presence of a thermodynamic economy that shapes national and natural history, guiding the formation of information, command and control structures that Adams believes will alter evolutionary outcomes. I will focus, in particular, on how Adams's fantasy of national belonging relies on race, both as a biological limit for citizenship and as an abstract force that animates the idea of women as *biological dynamos*.

I. Evolution as historical process

To contextualize Adams's interpretation of Darwin and Spencer, a discussion of how racial, class, national and gender discourses figure in evolutionary thinking in the early twentieth century will be helpful. Although the *Origin of Species* hints at the social implications of evolutionary theory, *The Descent of Man* contains Darwin's most direct comments on evolution and human behaviour.[9] His ambivalence about the direction of evolutionary development, however, resists any reading of his works as a scientific validation of a hierarchical view of race, nation, class and gender. Peter J. Bowler points out that the *Descent* exhibits a 'progressivist stance', marked by a pronounced embrace of Lamarckian principles, namely that acquired traits could be passed onto offspring.[10] Indeed, backing away from a doctrine of cut-throat competition, Darwin claims that sympathy for the unfit, as a moral value, connects to a larger effort to preserve the 'accumulation of capital', otherwise 'the arts could not progress; and it is chiefly through their power that the civilised races have extended, and are now everywhere extending their range, so as to take the place of the lower races'.[11] Yet his following disclaimer casts doubt on evolutionary progress: 'But development of all kinds depends on many concurrent favourable circumstances. Natural selection acts only tentatively. Individuals and races may have acquired certain indisputable advantages, and yet have perished from failing in other characters'.[12] Darwin's insistence on chance undercuts at key points an argument for continuous advancement through natural selection; even more, it disputes the cherished belief about the value of individual effort. In Darwin's model of evolutionary randomness, Bowler finds, 'there is no point in the individual's trying to work harder or improving its habits, since however hard it tries, rivals who are congenitally better equipped will beat it in the end. This is a philosophy of trial and error, a game of genetical Russian roulette, in which the chances of inheritance determine everyone's fate'.[13]

When Darwin does marshal natural selection to describe national progress, it does not line up with his views of gender difference. Consider his reading of evolutionary development from a transatlantic perspective:

> There is apparently much truth in the belief that the wonderful progress of the United States, as well as the character of the people, are the results of natural selection; for the more energetic, restless, and courageous men from all parts of Europe have emigrated during the last ten or twelve generations to that great country, and have there succeeded best.[14]

This is as strong of a case as Darwin will make for the progressive role of natural selection, downplayed by his later commentary on the mental development of men and women. Necessary for the masculine enterprise of nation and empire building, the qualities of genius, imagination and reason, he believes,

> have been developed in man, partly through sexual selection, – that is, through the contest of rival males, and partly through natural selection, – that is, from success in the general struggle for life; and as in both cases the struggle will have been during maturity, the characters gained will have been transmitted more fully to the male than to the female offspring.[15]

The inheritance of acquired characters becomes of greater importance, as the inequality of inheritance creates gender inequality. Yet despite their perceived physical and mental inferiority, women could be, 'when nearly adult, trained to energy and perseverance, and to have [their] reason and intelligence exercised to the highest point'.[16] But equality could only exist if those women who bettered themselves could reproduce, over 'many generations', in greater numbers than their less educated counterparts.[17] Not only does Darwin feel this outcome is unlikely; he also believes that men, who already have a head start, will continue to develop their minds through supporting themselves and their families, and hence perpetuate their apparent superiority.

Needless to say, although he reflects the biases of his times about race, class, civilization and gender, Darwin does not furnish any easy justification for them. That task is taken on by Spencer, whose wholesale acceptance of Lamarckian theory as the motor for evolutionary development argues that evolution equals continuous progress, plain and simple. Consequently his argument, Stephen Jay Gould explains, shows how a progressive view of evolution 'ranked human groups and cultures according to their assumed level of evolutionary attainment, with (not surprisingly) white Europeans at the top and people dwelling in their conquered colonies at the bottom'.[18] Even in this reductive form, however, evolution questions racial hierarchies through its transformation of biological development into a process whose logic fails always to validate the desired outcomes. Bowler elaborates: 'The great paradox of nineteenth-century social thought was that lower races were condemned to permanent inferiority, while the white race was supposed to be capable of continuous progress'.[19] Therefore Darwin and Spencer were often invoked together to explain biological development, as Darwin provided the scientific authority, while Spencer furnished the ideological weight. Yet such a practice tends to blur the differences between the two, and it is no small irony that, despite the 'comparatively rapid acceptance' of evolution in the middle to late nineteenth century, few had a more than limited understanding of how natural selection worked.[20]

It is not for nothing, then, that Adams comments, in the 'Darwinism' chapter, that natural selection was 'what a young man had to take on faith'.[21] That people could accept the broadest outlines of Darwin and Spencer's theories, and not recognize the distinct conflicts between the two, led to the grandiose rhetoric about uniformity and unity that Adams mocks throughout the chapter. Having read the *Origin* and reviewed the tenth edition of Charles Lyell's *Principles of Geology*, he was in a better position than most to address evolutionary debates and their misunderstanding. Adams, for instance, plays Lyell's belief in gradual change off against Clarence King, a geologist and Adams's friend, who advocates a more catastrophic model, giving, as J. C. Levenson puts it, the 'debate over Darwin a new turn'.[22] Laced with sarcasm, Adams's explanation for the appeal of evolutionary theory is that it offers an absolute set of scientific guidelines against which to measure and evaluate history:

> Natural Selection led back to Natural Evolution, and at last to Natural Uniformity. Unbroken Evolution under uniform conditions pleased every one – except curates and bishops; – it was the very best substitute for religion; a safe, conservative, practical, thoroughly Common-law deity. Such a working system for the universe suited a young man who had just helped to waste five or ten thousand million dollars and a million lives, more or less, to enforce unity and uniformity on people who objected to it; the idea was only too seductive in its perfection; it had the charm of art.[23]

There is a rather heavy cost to be paid for 'unity and uniformity': the bloodshed of historical conflict. As Levenson illustrates, Adams does not resist Darwin as much as he does Darwinism, that is, the name given to Spencer's popularized version of Darwin's theories; Spencer's claim that evolution held the promise of continuous human advancement enthralled a nineteenth-century popular audience.[24] Summed up by Adams as a move from scientific theory to more abstract principles, a shift from natural selection to natural uniformity idealizes scientific logic. But the replacement of religious order with natural order has more than a few discrepancies. One flaw that especially fascinates Adams, Levenson remarks, is the gap in the evolutionary scheme, represented by 'the ancestorless Pteraspis, which comes in as the first vertebrate (though it has since been demoted to a more primitive order of fish), and the do-nothing Terebratula, a primitive mollusk unchanged over successive geologic eras ... They were difficulties for Darwin, to be sure, but they pose insuperable questions for the simplified evolutionism of Spencer'.[25]

Adams's dismissal of evolutionary progress, I suggest, reveals an obsession about the loss of a continuous historical narrative, a 'Natural Uniformity' sustained by biologically stable formations of race, class and gender. His repeated insistence that history fractured at the year 1900 guides his search for a lost unity, the loss coinciding with the dispersion of national subjectivity as immigration and expansion pressure the defining limits of US citizenship. Invoked by the fracture at 1900, modernism, for Adams, is enlisted in the imperialist project through a social Darwinist construction of racial hierarchies. He writes: 'Society offered the profile of a long, straggling caravan, stretching loosely towards the prairies, its few score of leaders far in advance and its millions of immigrants, negroes, and Indians far in the rear, somewhere in archaic time'.[26] In a masculine fantasy of conquest, the caravan moves 'towards the prairies', setting the stage for imperialist desires to surpass continental boundaries. Yet those who lag behind, 'somewhere in archaic time', embody an evolutionary distance, a loss of unity forced by modernity and its disruption of national citizenship.

As Adams retreats from the supposed progress of the modern age, his feelings of disconnection intensify to the point where he classes his own subject position as outside the limits of citizenship (as an immigrant) and created out of historical loss (as Native American):

> Not a Polish Jew fresh from Warsaw or Cracow, – not a furtive Yacoob or Ysaac still reeking of the Ghetto, snarling a weird Yiddish to the officers of the customs, – but had a keener instinct, an intenser energy, and a freer hand than he, – American of Americans, with heaven knew how many Puritans and Patriots behind him, and an education that had cost a civil war. He made no complaint and found no fault with his time; he was no worse off than the Indians or the buffalo who had been ejected from their heritage by his own people; but he vehemently insisted that he was not himself at fault.[27]

Adams's juxtaposition of evolutionary instinct, thermodynamic energy and national identity (he is, after all, 'American of Americans') suggests that historical violence both creates and holds together the fractured subjectivity of the US citizen, crafted out of disparate identities and the drive for national expansion. Granted, this image is a fantasy of Adams's own sense of failure (highlighted by his 'education that had cost a civil war'), but it is animated by his performance of a historically doomed icon, the 'vanishing' Indian, described by Alan Trachtenberg as 'the idea that the indigenes of the continent would disappear as a matter of natural course'.[28] As US frontier conquest is replayed as the creation of an overseas empire, so too are scenarios of the vanishing Indian relocated. In his reading of Theodore Roosevelt's expansionist rhetoric, also discussed in Chapter 3, Richard Slotkin comments that 'the imperial project begins with the metaphoric extension of Frontier categories to a new situation in which Asians become figurative Apaches and the Philippines become a symbolic equivalent of [Daniel] Boone's Kentucky or [Sam] Houston's Texas'.[29] Informed by his earlier image of the caravan heading over the prairies, Adams's 'playing Indian' is staged at a key moment where Manifest Destiny becomes a global project.[30]

At this crucial juncture, Adams's performance also articulates the anxieties about the destabilizing of national subjectivity. If subjectivity, on the national stage, is a renunciation of the traditional role of individuals seeking and finding their heroic destinies, then we see how what Cindy Weinstein calls Adams's 'strategic self fashioning' as victim further enables this process of subject making.[31] Yet Trachtenberg notes that Adams's dual role as immigrant and native represents 'two imaginary terminals' that guide how '[f]earful and hostile responses to the sight of people who looked so "foreign" jamming the Great Hall of Ellis Island can be taken as the outer edge of a deeper anxiety: the inherently unsettled basis of national identity or nationality in this exemplary nation of immigrants'.[32] Adams's fantasy of the otherness of the immigrant ('still reeking of the Ghetto') connects with the vanishing Indian, who belonged to the land without having citizenship, to suggest that, as 'American of Americans', Adams feels a double displacement. But his declaration that he is 'not himself at fault' for the disappearance of the native inhabitants argues that history surpasses individual agency and displaces national responsibility for historical offences against the native inhabitants of the land. At the same time, his response signals a wilful misrecognition of the historical grounding of those subaltern identities he seeks to appropriate. His attempt to read tragic implications into national expansion reflects a case of 'imperialist nostalgia' (discussed also in Chapter 4) that, as Michael

L. Johnson observes, was felt by those who, having tamed the natural landscape of the Western territories, 'harbored an inauthentic longing to restore what they had destroyed'.[33] Thus Adams's feeling that he bears no responsibility for the erasure of Native Americans is a lament for the very conditions he has helped to create.

II. Thermodynamics and citizenship

Throughout Adams's narrative, financial and political upheavals that occur in the wake of economic progress question a gradualist model of evolutionary change. As Crosbie Smith and Ian Higginson comment,

> An increasing concentration of wealth and power in the hands of the few – the bankers, the capitalists, and ultimately perhaps J. P. Morgan – would be enough to suggest that catastrophism in some form would be a more realistic model for the New World disorder than that of gentle . . . uniformitarianism.[34]

Adams had a close encounter with the economic turmoil of the 1890s. As he tells us, his brothers' letters inform him of widespread bank failure, 'requesting his immediate return because the community was bankrupt and he was probably a beggar'.[35] While he was not ultimately affected by the crisis, it reinforced his view, as William Merrill Decker insists, that 'the geopolitical world that held for him an obsessive interest' appeared 'increasingly subjugated to the predatory design of international finance capital'.[36]

While Adams may have mixed feelings about progress, he rationalizes that capital might be harnessed through the technological innovation famously displayed at the 1893 Chicago Exposition. The highlight of the Exposition that Adams attended, the White City, was, in Slotkin's words, 'an architectural extravaganza in ersatz marble representing the pinnacle of Euro-American civilization'.[37] Shaping how Adams sees the Exposition as 'a step in evolution to startle Darwin', the White City allows Adams to place evolutionary theory within a racially guided process of cultural expansion on a transatlantic scale (the march of 'Euro-American civilization').[38] This investment in whiteness, if you will, fuels fantasies of national and global dominance through technological mastery.[39] In Chicago, he experiences the dynamo for the first time, announcing that it gives 'to history a new phase'.[40]

FIGURE 7 Westinghouse Exhibit, featuring dynamos, at the Chicago Columbian Exposition, 1893

Images of the dynamo (technological power) and the White City (racial power) thus impel him, Slotkin asserts, to conclude: 'A capitalist system had been adopted, and if it were to be run at all, it must be run by capital and capitalistic methods; for nothing could surpass the nonsensity of trying to run so complex and so concentrated a machine by southern and western farmers in grotesque alliance with city day-laborers'.[41]

Adams's deep ambivalence, however, about a transatlantic evolutionary system that is unable to hold, maintain or order race, class and gender signifiers motivates his turn to a thermodynamic model to preserve a national investment in whiteness. The chapter title 'Indian Summer' (preceded by 'Chicago' and his account of global travels, 'Silence') implies, according to John Carlos Rowe, that the White City, as the cultural twin of the dynamo, reanimates expansionism, 'that US Manifest Destiny is finding a new project'.[42] This investment is articulated through the linking of subject formation to a political agenda, expressed through the policies developed by John Hay, the Secretary of State and Adams's friend, that prioritize economic expansion and the policing of the Caribbean and Philippines. As discussed in Chapter 1, guided by an imperialist discourse that regards 'the Philippines as a question of balance of power in the east', Adams's well-rehearsed image of history breaking at 1900 augurs a new phase of American citizenship.[43] This new phase denotes a subject who 'had turned his back on the nineteenth century', capable of managing the forces of technological change.[44] The dynamo symbolizes the transition to a modernized capitalist economy, seen as the natural outcome of a thermodynamic refiguring of social organization and political control. Nearly a decade after the Chicago Exposition, thermodynamics makes a transatlantic crossing to the 1900 Paris Exposition. Here Adams speaks (in third person) of the dynamo as a force that will structure the chaotic processes of evolution: 'As he grew accustomed to the great gallery of machines, he began to feel the forty-foot dynamos as a moral force, much as the early Christians felt the Cross . . . [O]ne began to pray to it; inherited instinct taught the natural expression of man before silent and infinite force'.[45] Adams recognizes the dynamo as a focal point for the energies he sees compelling economic change, what Mark Seltzer calls 'the final giving way, in the 1890s, of a "simply industrial" economy of production to a "capitalist system" and "machine" ruled by the laws of conversion and exchange'.[46]

This capitalist 'machine', however, troubles traditional gender identities. Adams describes the new American woman as 'monthly-magazine-made', seeing her energies as debased.[47] He traces the purity of these energies back through history to the religious figure of the Virgin: 'She was Goddess because of her force: she was the animated dynamo; she was reproduction – the greatest and most mysterious of all energies; all she needed was to be fecund'.[48] Of course, as Weinstein points out, 'the woman's body in Adams's representation of perfect sexuality is actually the closest thing possible to the dynamo. Like the dynamo which creates force without sexuality, so too the virgin produces children and forces with little, if any help, from the body'.[49] The notion of perfect sexuality, expressed in thermodynamic language, impels Adams to declare that the Virgin 'having acted as the greatest force the western world had ever felt . . . the historian's business was to follow the track of the energy to find where it came from and where it went to; its complex source and shifting channels; its values, equivalents, conversions'.[50] Through Adams's attempt to chart the direction of the force-transformations that dissolve unity, thermodynamic ideas, broadly analogized, smooth out the contradictions among essentialist models of gender, reproduction without bodies and vitalistic conceptions of life. But he remains haunted, as Jennifer L. Fleissner observes, by the onset of a 'technocratic culture' and an antihumanist 'rationalized order' that reduce agency.[51] He claims that '[m]odern politics is, at bottom, a struggle not of men but of forces. The men become every year more and more creatures of force'.[52] Aligned with this loss of agency is, he notes, an overall thermodynamic dissipation that raises 'two large questions of vital importance to America: – inertia of race and inertia of sex'.[53] The two questions come together around the body of the American woman, viewed as a site for racial reproduction.

Adams tries various ways to engage the new American woman question, through a geopolitical analysis of Russia that casts racial discourses into pseudo-scientific images of 'race inertia', to the dynamo, which Seltzer remarks, is an 'extension' of the Virgin's thermodynamic life force.[54] Adams proposes that a larger historical direction, what he searches for throughout the narrative, could be discovered if a scientific theory of history could be developed. Yet his vision of the world as a thermodynamic machine only hastens the loss of gender subjectivity brought on by technological progress. He writes that the new American woman acts as if she were

> making a violent effort to follow the man, who had turned his mind and hand to mechanics. The typical American man had his hand on a lever and his eye on a curve in the road; his living depended on keeping up an average of forty miles an hour, tending always to become sixty, eighty or a hundred, and he could not admit emotions or anxieties or subconscious distractions, more than he could admit whiskey or drugs, without breaking his neck.[55]

The 'monthly-magazine-made American female' is paralleled by the American male's absorption into the machine. According to Fleissner, the American woman, for Adams, loses her sexual power: '[W]hen Dynamo has supplanted Virgin, feminism amounts to women's doomed attempt to follow men's lead and "marry machinery"'.[56] Adams's fantasy of thermodynamic control, however, applied to the new American woman, fails to distract us from the issue of race suicide that his thoughts so strongly embrace, as he argues that it is critical that the 'woman go on reproducing'.[57] Fleissner observes that Adams's adherence to the dominant strain of Roosevelt-era sexual politics forecloses the possibilities for the transformation of the American woman.[58] Thus he makes a circular argument for reproduction without bodies, a thermodynamic rewriting of the dictates of biological determinism.

Here Adams's scientific thinking becomes a form of miseducation. The chapter on 'The Grammar of Science' opens with Adams attempting once more to apply physics to solve questions of race and sex. Not only is his comparison of race and sex inertia to entropy in a closed thermodynamic system scientifically flawed; his metaphysical search for order through a unified theory of forces is halted by his reading of Karl Pearson's *The Grammar of Science* (1900):

> [T]he historian had no interest in the universal truth of Pearson's or Kelvin's or Newton's laws; he sought only their relative drift or direction, and Pearson went on to say that these conceptions must stop: 'Into the chaos beyond sense-impressions we cannot scientifically project them.' We cannot even infer them; 'In the chaos behind sensations, in the "beyond" of sense-impressions, we cannot infer necessity, order, or routine, for these are concepts formed by the mind of man on this side of sense-impressions'; but we must infer chaos: 'Briefly chaos is all that science can logically assert of the supersensuous.' The kinetic theory of gas is an assertion of ultimate chaos. In plain words, Chaos was the law of nature; Order was the dream of man.[59]

As Ronald E. Martin comments, Adams's 'hostile and sarcastic' response to Pearson underlines the futility of his quest and reveals his failure to grasp 'the very ideas that must be accepted if one is to be unsparingly educated'.[60] In his subsequent retreat into nostalgia, Adams juxtaposes the chaos of the present with the past, a unity guided by a metaphysical force, religion and a stable subjectivity constructed through the animating energies of the Virgin. Addressing the present, he declares: 'The child born in 1900 would, then, be born into a new world which would not be a unity but a multiple'.[61] Here Adams predicts the building up of systems perspectives through observing systems, that is, systems that observe other systems. If we regard Adams's inclusion of the historian in his own historical reading, then what we read is a self-conscious observation of history. Cary Wolfe's study of Niklas Luhmann's theories, as discussed throughout this book, sheds light on the 'epistemological' difficulties posed by and for Adams; Wolfe explains that Luhmann 'insists that all observations . . . are based on systems codes, which are themselves built on a constitutive distinction that is essentially paradoxical or tautological.

What this means . . . is that all such observations and systems are built on an inescapable "blind spot" that only *other* observations, from within *other* systems can reveal'.[62]

If we consider Adams's Virgin and Dynamo as a tautological construction of gender, the 'blindness' that sustains such a scientific–historical analogy, we can see Adams further struggle with a 'Dynamic Theory of History' that invents a scientific meta-narrative for history, making history a self-reflexive process: it does not unfold as much as it folds into itself through his frequent resort to paradox, tautology and contradictory logic. Martin points out the misapplication of and muddled thinking about science that clouds the Dynamic Theory, such as metaphors treated as logical sutures and postulates that have 'an element of casual arbitrariness'.[63] More acerbically, Ernest Samuels declares that his scientific theories about history either 'stand as a monument to intellectual ambition, or the most prodigious tale of a tub by which a supreme ironist defies reason to pursue him'.[64] And yet if Adams's Dynamic Theory is the ultimate symbol for the failure of a cohesive historical narrative, it discloses, as we shall see, how the US could be positioned at the forefront of global dominance.

III. The new American as techno-subject

Before we investigate the role of the US in running Adams's thermodynamic world machine, a brief reading of the Dynamic Theory is in order. For Adams, power structures act like fractals, reproducing their core values through social adaptation, whereby '[s]usceptibility to the highest forces is the highest genius'.[65] While humans struggle to use their evolutionary-honed capabilities to learn, a process shaped by the force of nature, they are ultimately unable to grasp something beyond their perceptions.[66] In 1900, there is 'the appearance of the new class of supersensual forces, before which the man of science stood at first as bewildered and helpless as, in the fourth century, a priest of Isis before the Cross of Christ'.[67] In sum, the plight of human education restages, on a larger scale, Adams's predicament.

In his role as historian, he depicts the writing of history as creating a tension between the displacement and re-emergence of agency. While the writing of history is not random, the pattern created is unstable: 'The pen works for itself, and acts like a hand, modelling the plastic material over and over again to the form that suits it best. The form is never arbitrary, but is a sort of growth like crystallization, as any artist knows too well; for often the pencil or pen runs into side-paths and shapelessness, loses its relations, stops or is bogged. Then it has to return on its trail, and recover, if it can, its line of force'.[68] Replicating the Dynamic Theory, the writing of history is a process of adaptation and reiteration, or 'crystallization'. The recovery of a 'line of force' connects the writing of history to the 'logical' outcome of political force that guides the Dynamic Theory, a connection that is reinforced when Adams defines writing as one of several 'economies of force' that construct history.[69] Although he argues that the observer constructs history, he does not realize the blind spots in his own observing, indicated by his lapse into a reductive social Darwinian scenario, where 'gunpowder killed whole races that lagged behind'; in Asia or Africa, there is even cultural regression.[70] Again, circular reasoning constitutes his argument, his inability to recognize that his use of evolution, as a scientific theory, justifies the same values that legitimate its explanation for the natural order of things: dominant cultural beliefs about nationality, race, class and gender. Adams extends this scheme to the thermodynamic realm, as the culmination of his scientific (mis)education is to write a précis of thermodynamic-modelled national policy that credits Hay as the architect of economic expansion.

And surely, such an outcome is what Adams realizes with his subsequent 'A Law of Acceleration', whereby the initial search for a 'new equilibrium' presages Hay's work for a global balance captained by the US.[71] As Rowe's skilful tracing of the imperialist undertones of Adams's narrative demonstrates, Hay 'represents the transfer of historical energy from one epoch – 1200–1900 – to the next – 1900–2000'.[72] In his elegy for Hay, Adams highlights Hay's application of thermodynamics to a transatlantic power alignment: 'He had brought the great Atlantic powers into a working system, and even Russia seemed about to be dragged into a combine of intelligent equilibrium'.[73] That Adams's prediction is founded on the same fuzzy logic as his scientific reasoning leads Rowe to explore how Adams's 'digressions' from observing the rise of US imperialism, rather than 'predicates of his *literary* modernism', are, in fact, 'discursive efforts to legitimate international negotiations conducted by Hay (and the McKinley and Roosevelt administrations), the details of which Adams knew quite intimately' that were 'made in the interests of a specific foreign policy intent on establishing a US imperium – not in lands, peoples, or even raw materials, but in the control of nations and, of course, their markets'.[74]

The indeterminacy that Adams discovers in any historical mode of observation, that which puts him at the modernist forefront, justifies the search for a political equilibrium and defers the question of who benefits from it. Indeed, such indeterminacy becomes a convenient way for Adams to avoid answering any political questions at all, as he repeatedly stresses the shifting contexts and multiple perspectives one would have to contend with to answer such questions.

Yet even if Adams's delays and digressions could be said to reveal a cynical relativism, there are certainly limits to his political manoeuvring. Rowe cautions against accepting 'the argument that the latter third of Adams's *Education* is a deliberate effort to distract readers from the new political power elite of men like Adams and Hay', for it 'borders on a reductive, conspiratorial, even *paranoid* approach to literature's ideological function'. After all, such an argument would depend on 'the widest possible readership for the book', when in reality it 'was privately printed, circulated among friends, and not distributed beyond that inner circle until after Adams's death'.[75] Instead, Rowe focuses on the parallels between the artistic imaginations of Adams and Hay. Rowe writes that Hay '"originated" nothing, merely playing with the historical fragments he acquired . . . In the balance-of-power politics that he played, Hay composed a new authority for the United States that bears comparison with that of the modern literary author's'.[76] So too do Adams's struggles with entropy sketch the artistic direction of modernism. Rowe suggests that 'for the modern artist, the entropic drift of the West could be reversed by way of unexpected combinations, new compositions of old elements, intellectual and disciplinary transgressions that would open the "closed system" required for the heat-death that is the end of thermodynamic entropy'.[77] Rowe then concludes that 'Adams's theorizing is performed in all good faith as the intellectual complement to the imaginative work of the new foreign policy represented by Hay'.[78]

While Rowe therefore argues for the structural similarities between the literary innovation of Adams and the political invention of Hay, such an argument hinges on the distinction between thermodynamically open and closed systems. Hayles clarifies that 'the premise that human evolution takes place within a closed system is incorrect. Life is possible only because our planet receives massive doses of energy from the sun everyday'.[79] This critical distinction is the historical basis for a shift in reading entropy (from evolutionary endpoint to informational complexity), which I shall now pursue to provide a more accurate reading of the 'paranoia' surrounding information, command and control structures in Adams's model of the national techno-subject. Adams starts the Dynamic Theory by asserting that humans are in a continuous feedback loop with nature: 'Two masses – nature and man – must go on, reacting upon each other, without stop'.[80] We can contextualize this opening statement through Hayles's account of attempts to improve thermodynamic efficiency: 'Feedback loops had long been exploited to increase the stability of mechanical systems, reaching a high level of development during the mid-to-late nineteenth century with the growing sophistication of steam engines and their accompanying control devices such as governors'.[81]

My rereading of Adams, as I will now elaborate, connects his speculations about how to channel new forms of energy to the development of cybernetics that Hayles traces. That feedback systems have repeatedly been used for social control is evident in Adams's view of religion as an economic investment. For Adams, religion is as sensitive to feedback as market forces: 'Constantine the Great speculated as audaciously as a modern stock-broker on values of which he knew at the utmost only the volume . . . he merged all uncertain forces into a single Trust, which he enormously overcapitalised, and forced on the market'.[82] As Howard Horwitz comments about the Dynamic Theory, 'phenomena are chaotic, without sequence, but at the same time recycle the same impulse and pattern'.[83] Feedback loops, as recurring patterns, connect nature, religion, energy and science. In Adams's account, the Cross becomes a 'symbol [that] represented the sum of nature – the Energy of modern science'.[84]

The corollary to the Dynamic Theory, the Law of Acceleration, anticipates a model of thermodynamically driven ecosystems. Adams refers to the sun as a force, while the key component of his calculations is coal power, stored energy from the sun that flows through global markets.[85] Yet Adams elevates politics over ecology (nature's economy fuels the national economy); for coal power guides the industrialization that makes the US a leader on the global stage and paves the way for a new idea of citizenship, the national techno-subject, the 'new American, – the child of incalculable coal-power, chemical power, electric power, and radiating energy, as well as of new forces yet undetermined'.[86]

Underwritten by thermodynamics, techno-subjectivity embraces key concepts in cybernetics. While the search, in foreign policy, for a 'new equilibrium', is complemented by the scientific effort to control 'the new forces' unleashed on society, such strategies prefigure the notion of homeostasis, in Hayles's words, the way that organisms 'maintain steady states'.[87] Feedback loops extend the idea of homeostasis to machines. Adams's theories about history constitute feedback loops as both informational and mechanical, anticipating the advent of cybernetics that for Hayles signals that 'three powerful actors – information, control, and communication – were now operating jointly to bring about an unprecedented synthesis of the organic and the mechanical'.[88] Considered as an early synthesis of organic and machine, Adams's techno-subject is the culmination of what Horwitz describes as a historical process of 'reflexive self-confirmation that constitutes human identity'.[89] Adams's self-conscious foregrounding of the observer in constructing history can be considered as an earlier representation of how discussions about reflexivity develop from an examination of how informational feedback loops operate. This examination, as Hayles chronicles in her reading of the transcripts of the post-World War II Macy conferences on cybernetics, is shaped through a rethinking of 'World War II engineering technologies that aimed to improve human performance by splicing humans into feedback loops with machines' and then proceeded to examine critically the image of 'an operator sandwiched between a radar-tracking device on one side and an anti-aircraft gun on the other'.[90] The discussions that took place rapidly complicated the idea of the operator as 'man in the middle', pointing to how an observer constructed this human figure.

And we might say, concerning the vexed politics of observation, so it is with Adams. His thermodynamic model demonstrates how historical feedback loops flow through religion to science and then to technology, the new American subject recursively produced by the machinery of national economic expansion. Yet his idea of feedback loops, abstract as they are, call attention to himself as an observer on the political scene. Adams thus seems troubled by a growing awareness of reflexivity, described by Hayles as 'the multiply encoded nature of language, which operated at once as an instrument that the speaker could use to communicate and as a reflexive mirror that revealed more than the speaker knew'.[91] From this angle, Adams's denial of his working with Hay is a desire to remain neutral, to stay above the fray, which is an obviously untenable position. But to paraphrase Luhmann, if Adams, as an observer, cannot see that he cannot see what he cannot see, then there are certainly significant questions about what Adams (in 'good faith') can and cannot see.[92] Such questions are expressed through Adams's earlier quoting of Pearson: 'In the chaos behind sensations, in the "beyond" of sense-impressions, we cannot infer necessity, order, or routine, for these are concepts formed by the mind of man on this side of sense-impressions'.[93] In his account of Hay's negotiations, Adams tries to remain an objective reporter and observer. Yet Adams's education is shaped through his self-invested portrait of Hay as a 'visionary' artist, who creates policy in an entropic void. Hay ushers in a modernist era of foreign policy that requires attention to all parts of the globe connected by feedback loops. Not only, then, could one region influence another; Hay's negotiations often had to address the multiple, and sometimes conflicting, interests and goals of those nations involved. That spheres of influence become connected and impinge upon one another impels Hay to devise a unified strategy for global management.[94]

Read through Adams, however, US foreign policy can be defined as an expanded version of Manifest Destiny that creates anxieties about totalization. As I have earlier noted, the Chicago stockyards in Pynchon's *Against the Day* haunt Adams's dreams of the dynamo, transforming his symbol of economic and national power into an image of 'unshaped freedom being rationalized into movement only in straight lines and at right angles and a progressive reduction of choices, until the final turn through the final gate that led to the killing-floor'.[95] While Adams struggles with a national will to power that supports a 'technocratic culture', he also fears an evolutionary escalation that outstrips human control:

> An immense volume of force had detached itself from the unknown universe of energy, while still vaster reservoirs, supposed to be infinite, steadily revealed themselves, attracting mankind with more compulsive course than all the Pontic Seas or Gods or Gold that ever existed, and feeling still less of retiring ebb.[96]

Attempting to deal with anxieties about both totalization and chaos, Adams's examination of policy making illustrates how influence shifts from the thermodynamics of military machinery to the realm of information management and control in diplomacy and foreign policy negotiations. As a result, Hayles explains that chaos becomes redefined 'as an energizing force capable of stimulating self-organization and increasing complexity'.[97] In Adams's account, Hay has no small part in establishing information as a political commodity, as overseas regions are mapped and researched. And of course, the end goal is to bring such regions under US influence, foregrounding flexible thinking and imaginative combinations of diplomacy and market control.

So the key is information, coupled with thermodynamic structures. Rapid industrialization has put the US in the global forefront, but it is only through control that this position can be sustained and extended. Adams recognizes this, but worries about what will create a balanced (not too much or too little) control, at the same time wondering whether such a balance, if achieved, could be sustained. His elegy for Hay marks a turn towards abstracted patterns of information. That is, Horowitz contends that the Dynamic Theory refuses, in an increasingly depopulated landscape, 'the units of measure that usually compose history, mainly human agency'.[98] Hay's vaunted Atlantic system, seen as a transatlantic system of information, will in this perspective outlast his physical demise; in Adams's speculative view, 'in his eight years of office [Hay] had solved nearly every old problem of American statesmanship, and had left little or nothing to annoy his successor'.[99] The elegy perhaps functions as a prelude to Hayles's story of how 'information lost its body'. She argues: 'The point is not only that abstracting information from a material base is an imaginary act but also, and more fundamentally, that conceiving of information as a thing separate from the medium instantiating it is a prior imaginary act that constructs a holistic phenomenon as an information/matter duality'.[100] But as so often seen in Adams's narrative, and especially in his encounter with Pearson's work, chaos, read as noise, threatens to disrupt this process.

There are two possible resolutions. The first is to conceive of systems that can operate far from equilibrium. Adams flirts with such an idea, but ultimately is too pessimistic. The second is to dream of a scheme of evolution that will give humans a new 'mind'.[101] Exhibiting the latter, the culmination of Adams's project is a meshing of evolutionary and thermodynamic systems, as the measurement of efficient energy use points natural selection in the direction of thermodynamic dictates. As discussed in Chapter 2, Eric D. Schneider and Dorion Sagan regard natural selection as making choices at a 'higher' level. Viewed as important units of selection, energy flows of ecosystems trace out the larger direction of evolution, 'the equilibrium-seeking organizations of open systems thermodynamics'.[102] A return to the image of the social Darwinist 'caravan', in the closing chapter of the *Education*, both reminds us of the imperialist energies that guide

Hay's diplomatic efforts and suggests that evolution must be refigured beyond Darwin's notion of gradual change: 'The new American must be either the child of the new forces or a chance sport of nature'.[103] Both culture and nature constitute the 'economies necessary for the education of the new American'.[104] Hence Adams's failed education combines (in imaginative and unexpected ways) the forces of information, control and communication that would emerge in cybernetics in the next half century.

IV. Beyond evolution: information, control and paranoia

The *Education* registers concern about the effects of techno-subjectivity, namely that as the body becomes interpenetrated by information systems, corporeal boundaries disappear. At the same time, Adams steers us towards a model of non-equilibrium thermodynamics (NET), where entropy becomes redefined as indeterminability. By following these two trajectories in Adams's thinking, we will see that, while he goes beyond a Darwinian evolutionary scheme, he cannot resolve his ambivalence about a global conception of citizenship.

Adams's deep investment in thermodynamics leads him to repeated images of increasing entropy as cultural decay and inertia. It allows him to substitute scientific limits for the socio-political constraints on his fantasies of race, sex, class and citizenship. But as my citation of Pynchon, often regarded as an observer of a history shaped by conspiracy, hints at, the question is to what degree are these fantasies paranoid? We should recall Rowe's dismissal of a paranoid reading of Adams as attempting to conceal his involvement in Hay's policy making. With regards to this dismissal, I agree. Yet I believe there is reason to interpret the paranoid tendency in a way directly opposed to Rowe's interpretation: it is not the absence of information (Adams's deliberate withholding of the evidence of his participation), but rather a reaction to fears about information overflowing the boundaries of the body that could be argued to display paranoia about the loss of racial, sexual, class and national markers of the self. Such a response is not simply an overreaction to scientific and/or cultural progress, as critics such as Martin have argued. Rather, this response is a genuine concern about how disembodied information argues against any notion of a stable self: if information can lose its body (as is suggested by Hay's transatlantic and global systems of information, control and communication that delineate a policy of imperial coercion), it predicts a mode of self-abstraction, Adams thinks, that may indeed be undesirable. In this light, Adams's search for a political equilibrium tries to affirm that the optimum condition of systems is that they function at equilibrium. Yet such an affirmation cannot be made, because systems can and do function far from equilibrium. Such an idea is supported by Schneider and Sagan, who consider life as an open, not closed, system; that nature abhors gradients (for example, temperature or pressure differences) becomes 'an unstoppable tendency where energy flow leads to different natural complex systems including life itself'.[105] In fact, a condition of equilibrium would bring about the maximum levels of entropy that Adams would try to forestall. This aspect of Adams's miseducation, his assumption that political equilibrium is synonymous with thermodynamic equilibrium, may go the furthest to bolster his paranoia.[106] That the fear shifts from being swallowed by mechanical systems to being absorbed by information systems telegraphs the very concerns that would trouble cybernetics in the late 1940s.

First, as Hayles explains, in 1929 entropy becomes linked to information through reinterpreting Maxwell's Demon, a creature imagined by James Clerk Maxwell in 1871 that could sort molecules, as proof that the 'more information there is, the less entropy; the more entropy is present, the less information'.[107] Claude Shannon refined this reading by proposing that 'the more unexpected (or random) a message, the more information it conveys'.[108] Hayles goes on to say that it is therefore possible that 'large entropy production could drive systems to increasing complexity'.[109] But her reading of Norbert Weiner's *Cybernetics* (1948) suggests that scientists retreat, in an act of what she calls 'intellectual celibacy', from scenarios where boundaries are radically redrawn. In her central example, Weiner's discussion about information and organizational hierarchies hinges on the 'analogy' of insect reproduction, as pheromones outside the body act in concert with hormones inside the body. This analogy, Hayles remarks, implies that the body could be regarded as 'a sort of permeable membrane through which hormonal information flows'.[110] She then explains that Weiner backs away from this 'disturbing' speculation, because it proposes that 'personal identity and autonomous will are merely illusions that mask the cybernetic reality'.[111]

If we return to Adams, his formulation of woman as biological dynamo constitutes a response to the idea of a 'permeable' body. Articulating a view of reproduction as imaginative as Weiner's, Adams imagines an act of production without bodies that creates a stable body and reconstitutes racial, sexual, class and national signifiers. Adams's Dynamic Theory, moreover, generates complexity as a response to entropy, a process that fits with a reworking of information theory. Through feedback loops, systems become self-referencing, which highlights the centrality of information to thermodynamic models of self-organization. Erich Jantsch remarks that 'a dissipative [i.e. thermodynamic] structure "knows" indeed what it has to import and export to maintain and renew itself. It needs nothing else but the reference to itself'.[112] That the relationship between thermodynamics and information has been traditionally expressed in terms of entropy – an idea introduced by Shannon and Warren Weaver in 1949 – can, however, be misleading. In statistical thermodynamics, Jeffrey S. Wicken explains, entropy refers not to disorder but rather indeterminability; he thus proposes to replace the concept of entropy in information theory with that of complexity.[113] Looking at the formal structure of Adams's narrative, Hayles finds a turn towards complexity through its embodiment of '*the process it describes in the dynamic theory of history*'.[114] Put slightly differently, according to Luhmann, in the process of handling complexity, systems differentiate themselves into subsystems, reintroducing the system/environment distinction: systems in turn become environments for their subsystems.[115] We could say that Adams treats the complexity of information, what his historical observations create, by developing a more complex relationship between himself as observer and the environment he observes.

It is worth noting that Adams's narrative displays facets of NET in more simplistic forms. He suggests, as we have seen, that following channels of energy is the key to understanding dynamic life and historical processes. He also indicates that energy flows degrade over time, albeit in the socially conservative estimation that the energies of the new American woman are a debased version of those of the Virgin. The new energy forms being discovered allude to the ways in which systems resist entropy. His failure to find unity, moreover, alludes to the idea (although it is one that he constantly resists) that systems operate far from equilibrium.

Adams's vision becomes a kind of creative destruction, as the old order is surpassed at 1900, for the thermodynamic energy of the dynamo, both as the motor of industry and the ideological force (channelled by Hay) that justifies US imperialism.[116] And in turn, the gendered national subject is transformed; the new American woman embodies race as both a biological limit for citizenship and as an abstract force that animates Adams's fantasy of biological dynamos. Suggested by Adams's claim that culture and nature form the 'economies necessary for the education of the new American', the Dynamic Theory is a scientific narrative about the triumph of capitalism, as it survives by adapting to change, by going with the thermodynamic flow.[117] Viewed in the abstract (as Adams would surely be inclined to do), capitalism, as Schneider and Sagan remark, can 'create new products (e.g., Henry Ford's automobile, the personal computer) even as it destroyed non-competitive industries (e.g., Ford's automobiles were ousted by General Motors, later "destroyed" by Japanese automotive production)'.[118] Thus Adams's search for a scientific education anticipates Ira Livingston's observation about 'the neoliberal ideological synthesis of the sciences'. As Livingston comments, 'it is now rather difficult to restore the sense of scandal in the resemblance between descriptions of self-organizing processes in biology and physics and the transnational neoliberalism they tend to underwrite'.[119] If there is a 'sense of scandal' in Adams's thinking, it surfaces, I propose, in his scenarios of information, command and control , which threaded through thermodynamic processes produce ambivalent results. He struggles with the realization that his proposal to control information systems through efficiency ratios represents the same logic – earlier seen in Hay's (trans)Atlantic system that attempts to maintain a state of geopolitical equilibrium – used to control the lives of newly globalized citizens.

This global turn renders explicit Adams's appraisal of Darwin's construction of evolutionary theory: that its mechanism of selection is too simplistic to explain the complex transformations of national subjectivity that in turn produce the new American. For Adams, evolution does not deliver on its promise of establishing an origin for national citizenship and identity. What overshadows his fantasies about Darwinian dynamos is the fear that evolution will naturalize and thus make permanent the condition of imperial homelessness he feels so intensely.

V. 'The Rule of Phase Applied to History'

Throughout our discussion of Adams's thinking, we notice that he tends to look for a stable location of reference outside history. If he understood history as a dynamic process, the understanding of which in part required an understanding of the observer's position, he only gradually came to a realization that science worked in a similar way. In 'The Rule of Phase Applied to History' (1909), Adams's scientific (mis)education continues as he attempts to read historical implications into Willard Gibbs's theory that the transformation of any substance – for example, a liquid changing into a gas – goes through stages of equilibrium, or 'phases'.

Here as elsewhere, the limits of Adams's scientific understanding are on display. The essay's premise is to find a scientific detour around what Adams sees as the constraint of self-reference that further complicates the problems of historical observation. Of course, inherent in this premise is the belief that scientific disciplines are *more* stable than those in the humanities. Adams must, then, have had a rather rude awakening when informed by Henry A. Bumstead, a professor of physics at Yale College who had read the first draft, that his initial claim that at the end of the nineteenth century 'chemistry was complete on its old atomic foundation' was misled. Bumstead told Adams that developments in 'atomic and molecular theories' not only meant that chemistry was still in disciplinary motion, but that Gibbs's phase rule was now being seen as fitting with, rather than antagonistic (as Adams had supposed) to these theories.[120]

Yet Bumstead did state that he was 'much inclined to believe the main thesis of the author that mathematical and speculative (hypothetical) methods are destined to play a great part in many other branches of knowledge besides the physical sciences'. For the moment, at least, Adams's disciplinary crossing appears to be validated. After feedback of this sort, Adams made several changes to what would result as the final version of the paper – especially adding passages that seemed designed to remind the reader that Adams was essentially functioning as a second-order observer and/or served as disclaimers by acknowledging the limits of Adams's scientific education. Regardless of the revisions, Martin correctly points out that the paper still reflects Adams's confusion about basic scientific principles, such as the difference between velocity and acceleration, leading Martin to conclude that 'the finished essay is clearly a product of the vaguely understood science, the facile analogizing, and the apocalyptic yearning that went into it'.[121]

Despite Martin's criticism, there are indeed important moments in the essay. First, Adams forecasts disciplinary conflict as he claims that scientific knowledge production, as guided by Gibbs's theory, predicts the end of history: 'this ethereal current of Thought' will eventually reach 'equilibrium in the ocean of ultimate solution'.[122] Self-reference for Adams becomes a hall of mirrors; each observation generates the next as a new phase or 'channel' develops, until the limit to transformation is reached. As far-fetched an idea as Adams's thought/ocean analogy is, it does resituate self-reference as a dynamic process, although it is only revealed in Adams's vision of 'annihilation' that forecloses meaning altogether.[123] Another way of putting it, however, would be to say that Adams's imagining of the end of history is *really* the end of historical meaning; that is, history is no longer an *exclusive*, or exclusionary, discipline. Again, that Adams's interdisciplinary perspective sounds in a negative (or even redundant) register is a given, we might argue, but through his fascination with the limits of thinking/meaning does open *all* disciplines to critical evaluation. When Adams, later in the essay, revisits evolutionary development, he does so from this perspective:

> [T]he whole phase of instinct survives, not merely as a force in actual evidence, but as the foundation of the whole geological record. As an immaterial force, Instinct was so strong as to overcome obstacles that Intellect has been helpless to affect. The bird, the beetle, the butterfly accomplished feats that still defy all the resources of human reason . . . The quality that developed the eye and the wing of the bee and the condor has no known equivalent in man.[124]

Aside from his summation of instinct as 'an immaterial force' (that, after all, has no scientific analogue), Adams here questions the primacy of consciousness, traditionally used to assert human superiority over non-human animals. What we see therefore is the emergence of Adams's distinctly posthuman stance (that will further be developed in 'A Letter to American Teachers of History'), albeit a position that does not escape the paradox of self-reference in cognitive science: if instinct is 'immaterial', it simply replaces consciousness at the top of the cognitive hierarchy of liberal humanist thought.[125] Regardless, Adams has expressed a posthumanist challenge to the ideal of unlimited human progress, for 'the physicist-historian' now faces the question, 'How long could he go on developing indefinite new phases in response to the occult attractions of an indefinitely extended universe?'[126] Adams goes on to confirm that 'thought is material', hence limited because it follows 'the law of gravitation'.[127] Cumulatively, in this respect, Adams's images of a comet and electricity[128] extend – through the device of repetition – his argument about the limits of a *materialized* human reason, reflecting his increasingly desperate attempts to push the limits of the disciplinary creation of historical observations through refashioning the principle of self-reference. More apocalyptic portents are to come, as we shall see.

In sum, the paper may be regarded as a multi-dimensional, interdisciplinary attempt to create a scientific (rather than a cultural) narrative of Adams's reaction to the dynamo, which anchors his contention (one with which Bumstead, among others, disagreed) that human thought fractured in 1900. If the dynamo does forecast a new technoscientific regime, then Adams does hint here at the multiple impressions that different academic disciplines might form of such an emerging regime. Perhaps, then, the key crisis to which the paper points is the inability to bridge the different disciplinary languages, a crisis which for Adams highlighted a mounting social problem: what are the ethical parameters of an early-twentieth-century education? And what answer would science, even if it could, give? Adams would attempt to answer these questions in his next thought-piece, 'A Letter to American Teachers of History'.

VI. 'A Letter to American Teachers of History'

In 'A Letter to American Teachers of History' (1910), Adams writes: 'As an Organism society has always been peculiarly subject to degradation of Energy, and alike the historians and the physicists invariably stretch Kelvin's law over all organized matter'.[129] Adams presumably got the idea of society as an organism from reading Hans Driesch's lectures at the University of Aberdeen (1907–8), compiled in two volumes as *The Science and Philosophy of the Organism* (1908). On a page of the second volume of his copy (housed at the Massachusetts Historical Society), Adams wrote,

> Does not this definition of organism apply most precisely to – The Church (Roman), An Army, (to wit, German), A State (e.g. England), A Federation (United States), A Corporation (US Steel), A Bee-hive, The role of Entelechy in an army is sufficiently self-evident. It is not the sum of its units. It *is* its coordination to an end.[130]

As Jane Bennett explains, Driesch regards entelechy as a 'self-directing awareness' responsible for the organism's capability for 'innovative action – it repairs injured parts, re-creates several ones, and adapts old parts to perform new roles – all to maintain the normal functioning of the whole and to preserve its identity'.[131] Entelechy, however, only applies to closed systems; as Bruce Clarke points out, 'the redescription of biological organisms as open systems' is a crucial development in systems thinking.[132]

Yet the notion of society as an organism lines up with systems theorist Niklas Luhmann's proposal that society is constituted through psychic and social systems. I want to argue that Adams's inquiry into the idea of social organization in the early twentieth century – and the pessimistic conclusion he draws from its being subject to the Second Law of Thermodynamics – sheds light on the emergence of complex systems, a signifier, for Adams, of a modernity that initiates the collapse of American empire. Moving beyond the idea that the 'Letter' is a final summation of Adams's melancholic withdrawal from the world, I will show how it clarifies his thinking about techno-science, history and the fate of the national subject.[133]

i. 'The Last Man' at the end of history

The Second Law's prediction of chaos suits Adams's pessimistic worldview, allowing him to align any and all historical indicators with his doomed vision – for instance, according to his logic, the acceleration of modernity will, perforce, increase 'inertia' – and counters more optimistic scenarios of social progress. Adams posits that 'society may be likened to an island surrounded by a rising ocean which silently floods its defences. One after another the defences have been abandoned, and society has climbed to higher ground supposed to be out of danger'.[134]

Perhaps here Adams is rethinking his earlier claim, in the *Education,* about the legacy of his friend, Hay: 'He had brought the great Atlantic powers into a working system'.[135] As Paul A. Bové elaborates, Hay's foreign policy goals support the (ongoing) American System, the name given to the historic tendency of US foreign-policy objectives to focus on global hegemony.[136] Adams thinks this system, alluded to as a 'rising ocean', is running out of control, Bové explains, because 'American intelligence lags American force – this is an Adams truism that typifies the peculiar danger the United States poses to the world and to itself and its own residual originary democratic ideals'.[137] With the United States unable to marshal its intellectual reserves to manage a crisis its imperial ambitions have ironically helped to catalyse, there is an absence of global leadership that might bring order to the chaos of modern history.

But taken to its logical end, Adams's argument becomes paradoxical. As William Jordy comments, Adams would have us believe that 'everything was best before anything had happened'.[138] Thus Adams rewrites nostalgia as a projection for a cultural order impossible to maintain, himself performing the role of 'last man' to witness the end of history.

The foreseen failure of the American System is transposed, through the 'law of degradation', to the psychic component of identity. Restating on a scientific level his truism about the primacy of 'force' over 'intelligence', Adams claims that 'in the dissipation of the higher intensities, a sympathetic exhaustion must be expected in all the energies dependent on the central system'.[139] What we have hence is the unravelling of a psychic life that constitutes the idea of subjectivity, foreclosing the production of the new American. Adams, moreover, amplifies the concern for how the American System encapsulates the danger the nation poses to itself into the spectre of self-destruction. Acting the role of second-order observer (his well-honed stance as a historian who observes observations), he quotes from Emile Durkheim, a prominent sociologist of the time, on suicide: '[T]he pessimist current has reached an abnormal degree of intensity due to some perturbation of the social organism' and reflects a 'hatred and disgust' for the future, a desire for 'destroying the real, and escaping from it'.[140]

ii. The paranoid style in American systems theory

As the Atlantic System (which can be conceptualized as a subsystem of the American System) runs out of control, Adams concludes that, properly taught, 'American history' would self-reflexively chronicle '"a universal tendency to the dissipation of energy" and degradation of thought'.[141] He pronounces that psychic life, the basis for consciousness and the source of self-identity, is already degraded, since 'Reason can only be another phase of the energy earlier known as Instinct or Intuition', and ultimately, 'the intrinsic force of Reason . . . must submit to the final and fundamental necessity of Degradation'.[142]

Although Adams seems resolute about the end of history, the national subject and US global empire, he discloses another perspective to his argument voiced as a second-order observation through the form of the letter to the addressee (American history teachers). As Vern Wagner remarks, Adams is asking his readers if they have any idea what they are teaching.[143] His attempt at irony, whether deliberate or not, may be 'feeble' according to Jordy,[144] but he raises some significant questions: how can this 'new' history be taught, and how then are disciplinary barriers redrawn in the light of how reason itself is shaped and guided by something *beyond* itself. These questions bridge the two parts of the 'Letter'.

That reason is not regarded as autonomous (that is, beyond the pull of thermodynamics) marks a posthuman turn in Adams's thinking, here delivered in a much stronger register than in 'The Rule of Phase Applied to History'. Located within his repeated assertion that a thermodynamic doctrine delivers an edict of endless and unavoidable degradation is the idea that human identity is *not* manifested through an act of intention (staged from a transcendental position), what Cary Wolfe calls the trope of 'the disembodied Cartesian subject'.[145] Wolfe goes on to assert that a serious review of Cartesian dogma would extend the heretofore exclusive (to humans) trait of consciousness to non-human animals, hence eroding the supposed difference between human and non-human animals. Adams, while he articulates this trope of posthumanism, focuses on the debate between the cases for 'conservation' and 'degradation' of energy – indeed, from his point of view, he is making the case for the primacy of the Second Law by stating the destructive effects of humans on the natural environment, one shared by all life forms: 'Indeed, the evolutionist himself has complained, and is still complaining in accents which grow shriller every day, that man does more to dissipate and waste nature's economies than all the rest of animal or vegetable life have ever done to save them'.[146]

In part, through Adams's reference to 'nature's economies', the modern (posthuman) nation-state comes to be seen as an assemblage of depleted natural and cultural resources. As a global extension of the nation-state, the American System, Adams reminds us, is an ineffectual attempt at maintaining social order – a geopolitical arrangement rapidly fragmenting, as we will now discuss.

As he relates in an (in)famous chapter in the *Education*, the dynamo announces a new technoscientific regime which intensified Adams's belief that modernity created a historical rupture. This regime is a 'virtual' world, in so far as subjectivity seems born out of a critical loss: the foreclosure on the making of the new American. On the crucial matter of how virtuality mediates the conflict between absence and presence, we may turn to Jacques Derrida, who considers the question of 'a politics of the virtual', of 'a certain delocalizing virtualization of the space of communication, discussion, publication, archivization', against that of whether we can merge what are regarded, currently, as two fundamentally opposed concepts: 'the thinking of the event to the thinking of the machine'.[147] The notion of the event is a sense of presence, while the machine is a sign of absence, because it follows a 'program without affect or auto-affection'. But if the two could be combined, Derrida continues, '*[N]ot only* . . . will one have produced a new logic, an unheard-of conceptual form . . . [T]his new figure would resemble a monster'.[148]

Wolfe calls systems theory, traces of such thinking that the 'Letter' displays, just this sort of 'monster' (a merging of event and machine), for there is no discrete event that produces a system; a system produces itself (it is, in systems-theory parlance, autopoietic), a self-production that in Dietrich Schwanitz's words 'is anything but mechanistic'.[149] For as Wolfe states earlier, systems theory's insistence on the separation of psychic (consciousness) and social systems (communication) makes good on Derrida's 'deconstruction of the . . . voice-as-presence'.[150] Wolfe later explains that, if it were not for such a separation, 'how could any psychic system maintain its own functions if it shared its unity with other minds? How could I deliver a lecture if I shared the moment-to-moment ebb and flow of psychic activity of even one other consciousness in the room?'[151]

The *monstrous* system of techno-science derives from the same thinking that explains psychic and social systems. Adams, however, pictures the birth of techno-science (as a signifier of modernity) as rather disturbing, a sort of mutated post-empire American System running out of control. Faced with the consequences of this creation, he does recognize that ethics cannot surmount the boundaries of a system – that is, it must develop within a specific system, not beyond it. What Luhmann calls functional differentiation (whereby, under the regime of modernity, systems respond to environmental complexity by creating subsystems that replicate the system/environment distinction), as Wolfe explains, 'guarantees that the subsystems can carry out their functions so that, for example, a legal decision is based not on the code of knowledge/ignorance (education) or profitable/not profitable (economic) . . . but rather on the code legal/illegal'.[152] Likewise in the 'Letter', the operational code for one academic discipline, such as physics, is incompatible with another, such as history. The ethical concerns for history cannot therefore be shared by physics. As Wolfe further explains, the attempt to surpass disciplinary 'closure' informs the objective perspective of the humanist model. This humanist order would be questioned by what Wolfe calls 'transdisciplinary' thinking, a subjective mindset that preserves disciplinary differences. It constitutes 'a kind of distributed reflexivity' guided 'by the fact that (by definition) *no* discourse, no discipline, can make transparent the conditions of its own observations'.[153]

While this posthuman idea is what Adams has been expressing all along, his evocation of society as an organism belies *the paranoid style in American systems theory* by picturing family structures as dissolved, human identity/agency subsumed by the system: '[T]he social organism, in the recent views of history, is the cause, creator, and end of the Man [*sic*], who exists as only a passing representative of it, without rights or functions except what it imposes'.[154] His conclusion returns to a nostalgic projection of the past, for the only 'complete solution' to the problems he has been describing 'seems, to an impartial bystander, to call for the aid of another Newton'.[155]

The paranoid style mentioned above is articulated through Adams's coupling of thermodynamic degradation with moral decay. As William Rasch details, a rather disturbing aspect of modernity, according to Luhmann, is that morality 'has detached itself from its premodern locus in religion and has become a self-replicating, parasitic invader of the various modern, functionally-differentiated social systems'. Consequently morality infects systems codes so that, rather than reflecting the needs and goals of specific systems, codes govern the operation of systems based on the 'binding' determinant of '"good" (or "bad")'.[156] In response, ethics 'emerges as the by-product of a system's attempt to preserve its own reproduction from the ravagings of moral infection'. Thus moral decisions are rewritten as a 'consideration of action in the face of competing alternatives'.[157] Wolfe comments: 'In this way, ethics attempts to ensure the autopoiesis and autonomy of the various function systems by limiting, and, if one likes, relativizing morality'.[158]

The major conflict in Adams's thinking is hence between his establishment of limiting, selective, differentiated ethical disciplinary parameters (so that, for instance, history cannot morally advise physics) and an insistence on a far broader moral code, the decay of which fuels Adams's pessimism turned paranoia. To put it bluntly, Adams misreads science by pushing its meanings beyond disciplinary constraints to claim that thermodynamic degradation is a universal law with moral dimensions. As Rasch would agree, that morality acts as a parasite on Adams's scientific speculations derails the logic of Adams's argument. At the end, Adams, who I believe – in a further ambiguity – could be naming himself as 'the impartial bystander', claims for observation to have an unquestioned objectivity, a claim that systems theory (as we have seen) will not validate. What then helps to fuel Adams's paranoia is a fear of closed systems, not recognizing of course that his own paranoia (paradoxically) is a closed system.

At first glance, the 'Letter' displays how paranoia becomes, for Adams, 'noise' in the system, a disturbance that resists any attempts at (self-)regulation. But in a systems-theoretical context such a reading overstates the case of noise, for to a second-order observer noise can be a sign of increasing complexity in a psychic/social system (as discussed earlier in my reading of the *Education*). If a system can deal with noise through a contingent process of dealing with environmental perturbations, paranoia conversely becomes, when interpreting the 'Letter', a staging of how, according to Luhmann, morality parasites off of the self-maintenance of social and psychic systems. Masquerading as a *final* order (construed as a location beyond system boundaries), morality sets itself up as an arbitrary standard to which the noise processing of systems conforms.[159]

To return to the conflict between morality and ethics in the 'Letter', morality, in my view, falsely supports the reading of noise as degradation, chaos, increasing entropy, a reading that overrides ethical concerns in/for specific systems – for instance, the ethical costs of 'the burnt-out energies of empire', manifested by/through global systems.[160] Perhaps Adams's attempts at such grim calculations cause him to realize, as Bové suggests, that 'the system cannot be made to work', an observation voiced in a 5 July 1905 letter to Elizabeth Cameron that forecasts 'the crisis of the 1914 European war and the concomitant social upheavals'.[161] Adams now feels like an empty subject, hollowed out by the end of empire.

CONCLUSION

We return to the US foreign policy refraction of the dynamo: the American System. From Paul A. Bové's perspective, Adams's historical writing chronicles 'the continuity of the American System . . . from the War of 1812 to the period between the Spanish–American War of 1898 and the onset of the European war in July 1914'.[1] Threaded through our discussion of US foreign policy as a truly modern signifier, for Adams, is the question of whether such policy constitutes a repackaging of colonialism. That is, Hay certainly is transforming foreign policy into a growth economy underlined by what John Carlos Rowe describes as a move away from 'territorial or even spatial domination' and towards 'the command of "representational" power either in commercial markets or their equivalents in cultural productions'.[2] Rowe, however, insists that Adams's response to modernity marks a new phase of colonialism: 'Adams advises the most powerful leaders of his age to pursue policies that would consolidate the twentieth-century economic authority of the United States in international trade and serve as the [basis] for that special brand of US *neocolonialism* understood in the era of the Vietnam War in the phrase "establishing spheres of influence"'.[3]

Thus a way to read Adams, Wharton, Dreiser and London (who will appear at the end of this discussion) together is how they react against and respond to this shift in US foreign policy – as they stage in different ways (through categories of race, class, gender and citizenship) a naturalistic projection of national imaginings of 'spheres of influence' as the biopolitical control of life – and the cumulative ecological consequences.

I. Henry Adams: ecocritic?

As we have said, Adams's thinking about the American System restages his vision of modernity which frames a shifting world picture as a scene of collapse, brought about by loss of control and the dissipation of national identity. Bové points out that Adams, in his 9 July 1905 letter to Elizabeth Cameron, 'explains the geopolitical ambitions of the American System as a plan to dominate, first, the North Atlantic world, from France to Russia, and then to "run our system through Asia" so that, in fifty years, the centre of the system would be in China, "the coal-fields of Shan-hsi [*sic*]"'.[4] Such ambition only magnifies the physical absence of a commanding structure, for Adams labels President Roosevelt as 'pure act':[5] a national performance that serves to mask the problem with the American System – albeit with global consequences – that Adams insists the nation must face. As Bové sees it, the American System, for Adams, is essentially an expanded version of the 'American capital system' that rests on a 'central and dangerous contradiction'. Bové explains:

> On the one hand, the surplus of power results from forces acting always in advance of those who try to understand or guide them, while on the other hand, those same forces produce a population of consumers and elites predominantly unreal in their ignorant dance within the world these forces compel into being.[6]

Given that, as Bové earlier confirms, 'Adams always wrote history from the point of view not of his own time but of what he could best understand about the emerging future ascertainable dominant forces seemed likely to produce',[7] Adams traces a shadowy regime of capitalist acquisition that haunts our current popular and scientific debates about technoscientific empire and the remaking of selfhood. As we know, Adams, in *The Education of Henry Adams,* tied the forging of a new national identity to technological mastery over energy resources to be achieved by the 'new American – the child of incalculable coal-power, chemical power, electric power, and radiating energy, as well as of new forces yet undetermined'.[8] Reading Adams, we see that political problems with the American System – earlier summed up by Bové as the idea that 'American intelligence lags American force'[9] – have, in turn, given rise to ecological problems. For example, in 'A Letter to American Teachers of History', Adams escalates his warning about how the American System as a growth economy is susceptible to natural perturbations, so that the 'flooded market' is the counterpart of the flood of 'surplus of power'.[10] Adams further laments, in the 'Letter', the human tendency 'to dissipate and waste nature's economies'.[11] He forecasts how pollution from coal and chemical waste threatens the global ecosystem, the cost of unregulated economic growth that has propelled the nation to the forefront of the world stage. Yet its most formidable economic competitor (as Adams predicted), China, has an even worse environmental record, displaying how the ecological consequences of modernization, enabled by the use of coal power and lax regulations on industry, seem to have accelerated.[12]

In addition, excessive consumerism hastens environmental destruction, and as Stacy Alaimo notes, mass-media gatekeepers validate 'the pervasive sense of disconnection that casts "environmental issues" as containable, eccentric, dismissible topics'.[13] Acting against a faster and more widespread recognition of ecological crisis has serious consequences, for when Adams refers to 'new forces yet undetermined',[14] he might well be predicting the (still not fully 'containable') dangers of nuclear energy, its potential for catastrophe evident in the global mapping of the disasters of Three Mile Island, Chernobyl, and Fukushima (caused, in part, by the location of the reactors that allowed for an increased risk of flooding posed by the natural factors of geology and oceanography) in the late twentieth and early twenty-first century. So what Adams would say are the political problems with the American System – that power outstrips knowledge – resonate with ecological problems on overlapping local/global levels, such as the disparity between the widespread use of chemicals and a complete understanding of their interrelated effects on humans, non-human animals and the shared environment.

Adams, in addition, anticipates the theorizing of global eco-politics. As we have seen, what drives Adams's vision of chaos, at every social/scientific level, is a sense of modernity as a traumatic fracture with the past. The notion that modern times signify social transformation at a fundamental level is shared by Ulrich Beck and Niklas Luhmann. Beck's concept of risk is central to how he defines modernity, much as systems theorist Luhmann's insistence on contingency is a vital part of his conception of modernity. The fact that for Beck a risk society is an effect of '"reflexive modernization," in which modernizing processes transform not traditional social structures, but those created by earlier waves of modernization' suggests an instability at the modernist core.[15] Viewed as a reiterative process, 'reflexive modernization' resonates with Luhmann's idea that, as William Rasch puts it, 'modernity is necessarily contingent . . . there is no escape from partial and conflicting perspectives, no escape from the vertigo of ever-shifting self-descriptions'.[16]

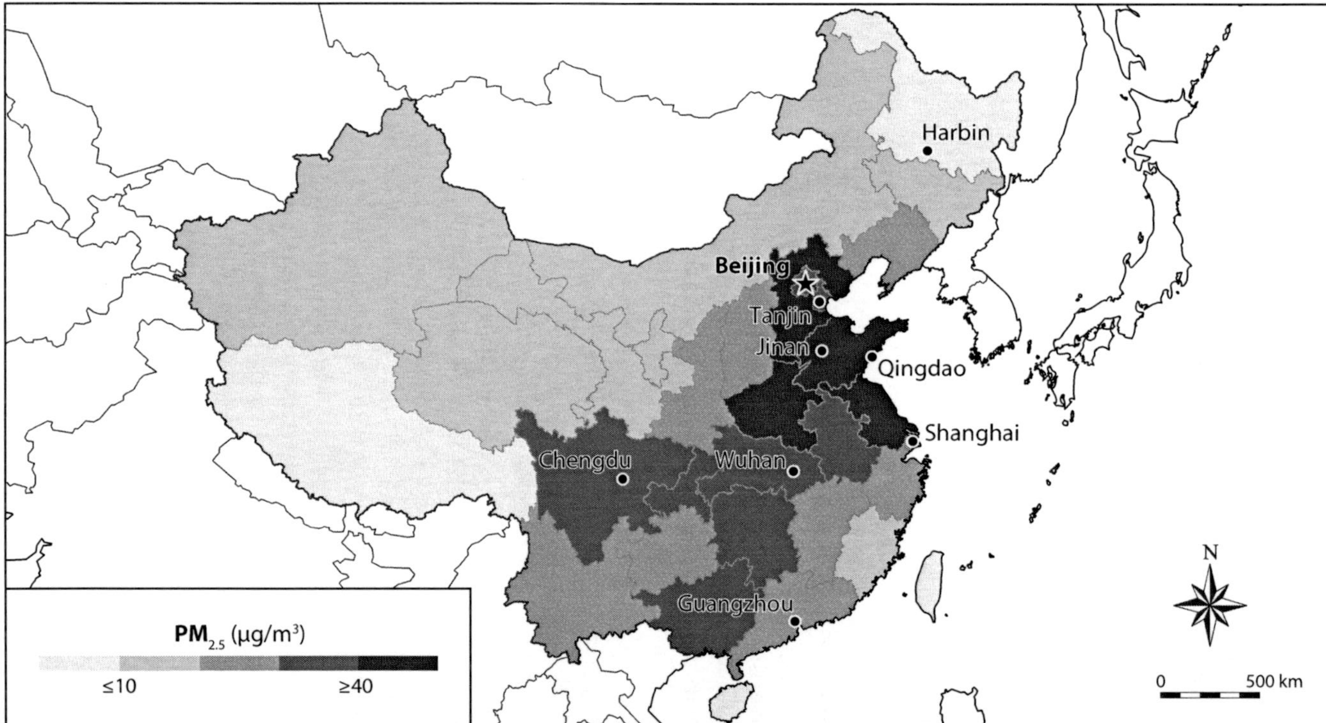

FIGURE 8 Airborne pollution over China, 2008–10 (source: NASA)

Beck's model of a risk society, which emphasizes the 'unintended "side effects" of modern technology',[17] is also 'necessarily contingent' because he thinks that 'ultimately no one can know of risks if knowing means having consciously experienced them'.[18] What Ursula K. Heise calls 'deterritorialization', which references 'the detachment of cultural practices from their anchoring in place', would seem to encapsulate Luhmann's modernity, described by Rasch:

> The 'fragmentation' of modernity has led to a 'fragmentation' of observation, which leaves us with no access to a commonly assumed objective world. What we see, in other words, is not a single elephant described by various blind men but various elephants made visible by different blind spots; and this modern 'loss' of objectivity cannot be compensated for by the 'rational' demand to see what we cannot see.[19]

In short, I would argue that Luhmann gives us a systems-based reading of Beck's risk society by describing how systems react to environmental complexity. According to Luhmann, 'modernity is characterized by the differentiation of autonomous systems'.[20] As we have seen previously, these systems, to handle complexity, differentiate themselves into subsystems, reintroducing the system/environment distinction: systems in turn become environments for their subsystems. As Luhmann points out, the first priority for systems is to maintain themselves, what is known as autopoiesis, even while they are environmentally sensitive to the degree that 'systems pursuing a trend toward exposure to ecological self-endangerment are eliminated'.[21] Rasch adds that a characteristic of modernity is that '[n]o system can take over the functions of any other system, nor can a system subordinate the functioning of any other system to its own'.[22] The ecological consequences are not hard to see. Andrew McMurry remarks that if all systems 'maintain autopoiesis without regard to one another', then the economy, 'which is fundamentally a system of exchanging anything for anything else for profit', will do anything for monetary gain, including damaging local and global environments.[23]

While Luhmann is pessimistic about systems recognizing and responding to environmental threats before it is too late, he does say that 'to explore how a society can react to the exposure to ecological dangers the constraints on the possibilities of its subsystems must be examined'.[24] Following Luhmann's logic, McMurry speculates how the destructiveness of the economy could be halted:

> [O]ne might reasonably wish for a day when the economy would be impotent to pollute and degrade the planet's ecosystemic services or draw down its unrenewable resources because to do so would instantly disturb and threaten the operations of the systems of health, science, law, and so on, which would now be as complexly single-minded as the economy. These other systems would consider any deployment of the old economic program (that is, processing all nature as externality or fodder) for what it has always been: a mortal threat to their own autopoiesis. Before the economy could do its dirty work these other subsystems would have sounded a klaxon that the political system could not ignore.[25]

Such an environmental alarm, I think, could be spelled out as shared or collective risk. Risk, like contingency, which must remain finally unknown, does not offer access to an unmediated, objective 'premodern' reality. Rather as Heise reminds us, what we could say is that the contingent knowledge of and about risk, the 'logic of "secondhand experience" and "secondhand nonexperience" can also be expected fundamentally to transform modes of spatial belonging and inhabitation'.[26] Such a transformation could animate McMurry's hope that social systems may be reconfigured.

Conversely, Adams's meditation on the chaotic functioning of the American System reminds us of the limits modernity (as emerging complexity) places on observation as a means for guiding (ecological) problem-solving. Citing Adams's 26 June 1905 letter to Cameron, Bové comments on Adams's 'seeing the world as dream image', which 'causes him nothing but near despair'.[27] Perhaps now Adams is realizing the narrowing of distance caused by modernity/globalization (he is stating his observations on the national future from Paris), and yet his thoughts seem to illustrate a rethinking of the dynamo. He realizes what he was unable to see before, namely the Pynchonian scenario discussed in Chapter 5, whereby 'unshaped freedom' became 'rationalized into movement only in straight lines and at right angles and a progressive reduction of choices'.[28] In the letter to Cameron, Adams exclaims that 'the constant narrowing of the horizon all round us is at times unendurable'.[29] What interests me is how Adams's 'near despair', that so often escalates into paranoia, emerges from his understanding of the abject failure of the Atlantic System – reminding us of the limits of systems-theoretical thinking (following Luhmann) for addressing what Bové calls 'often-unreal dilemmas'.[30]

Here through Adams, we can gauge the divergent pictures that Beck and Luhmann paint of a postmodern ecological future. Beck's risk society, Heise remarks, is not regarded as 'a fully established social pattern, but as an emergent one that at the moment overlaps with the structures of the modernist scarcity society',[31] just as the functional differentiation of social systems, from Luhmann's perspective, is constituted through the contingency of modernity. But if Beck and Luhmann line up in this way, in another way they do not. For Beck predicts, in Heise's words, 'the disintegration of the capitalist class society and of modernist approaches to knowledge'.[32] While Beck envisions a condition beyond modernity, Rasch notes, in Luhmann's view, that economically impoverished regions shaped by social inequalities demonstrate that '[w]hat modernity cannot "modernize" returns as a violent spasm, indicating that even functional differentiation, the great ethos of development, has its other and its limit . . . Modernity's other, it turns out, is neither "pre-" nor "post-"'.[33] Thus Adams's probing discussion of the failure of the American System (a grim reminder of the conflict between systems *theory* and *practice*, we might add) prepares us to see how Luhmann's scepticism about reformist thinking under the regime of modernity places, I suggest, a useful limit on Beck's project. As Heise has commented, in Beck's rather utopian thinking, '[f]ar from intensifying social inequalities, global risk scenarios will gradually undermine them'.[34]

Perhaps the limits of development (glimpsed above) connect with Adams's return of the repressed in his thinking about modernity: those races consigned to oblivion or extinction by modernization. It then follows that the appearance of *marked bodies* (by race/gender/class), produced through the modern regime, in the American System supports Adams's gloomy historical predictions about informational and social entropy as a signifier of modernity. These predictions are disclosed in two scenes in the *Education*, previously discussed in Chapter 5, the social Darwinist scenarios of 'a long, straggling caravan' and gunpowder as an agent of racial extermination.[35] In particular, Adams's white, male nostalgia for a unified state surfaces in his (in)famous systems-theoretical nightmare in the 'Letter': '[T]he social organism, in the recent views of history, is the cause, creator, and end of the Man [*sic*], who exists as only a passing representation of it, without rights or functions except what it imposes'.[36] Such a scene, as we shall see, reflects the *bad-faith* logic of his (female) reproduction without bodies, that is, a failure to think completely through the issue of the gendered subject.

II. 'Cyborg politics' and the technoscientific regime

Bové writes that Hay's innovations, such as the 'Open Door' policy with China, stage US global dominance, for no competing power could 'rival the American System in its ability to innovate, accumulate, dominate and modernize'. Indeed, the American System announces a modern turn in history, as it was 'more advanced and modern than any previous system. In effect the United States made outmoded all preceding sets of relations and the imbalance of power proved it'.[37] In the *Education*, however, Adams is made uneasy by the new American woman as a gendered reflection of Hay's modern strategy for the control of cultural productions. Adams describes the new American woman as 'monthly-magazine-made', seeing her debased energies as a sign of thermodynamic degradation.[38] Finding a metaphysical purity in the pre-modern religious figure of the Virgin, Adams suggests that the 'greatest and most mysterious of all energies' for racial reproduction bridges an otherwise insurmountable historical chasm. Cindy Weinstein remarks: 'Like the dynamo which creates force without sexuality, so too the virgin produces children and forces with little, if any help, from the body'.[39] While, for Adams, historical chaos brought on by thermodynamic decay takes on a gendered form, more crucially, his misreading of entropy limits his ability to think through completely the issue of the gendered subject.

Indeed, the 'new woman' becomes the prevailing model for national techno-subjectivity, a product of a conceptual shift from closed to open systems. Adams overlooks this important shift in systems thinking, whereas Hay's transformation of foreign policy appears to reflect its influence. According to Rowe, Hay, as a prime architect of modern cultural production, understands how 'the entropic drift of the West could be reversed by way of unexpected combinations, new compositions of old elements, intellectual and disciplinary transgressions that would open the closed system required for the heat-death that is the end of thermodynamic entropy'.[40] That open systems reverse entropy by using outside energy sources guides the emergence of the protagonist, in Dreiser's *Sister Carrie*, as a cyborg, a manifestation of Hay's (self)-representational power.[41] Carrie embodies the principle of self-organization that, in a modern era of open thermodynamic systems, characterizes all global life processes. Ira Livingston comments that 'we come to *recognize* self-organization because we have *attained* it to some critical degree. The common term for this process is simply *globalization*'.[42] As addressed in Chapter 2, a crucial thermodynamic trope, *surfing*, can be read into the scenes of Carrie's rocking: life becomes a 'process which has emerged on top of the flows of energy and cycles of materials that characterize . . . open systems. Life perhaps might be seen to have evolved in a way to ride these flows and cycles the way a surfer rides the flow of energy in water waves'.[43]

The circular logic of globalization, in turn, animates communication/information systems that push beyond the local. Carrie's fame as a New York actress is constituted through the interconnections between the stage-star system and the media. Borrowing from Niklas Luhmann's phrasing, these two systems are functionally differentiated; they are separate and yet interpenetrate one another through the language of celebrity, a connection most visibly represented in Carrie's name 'blazing in incandescent fire' on a neon sign at the theatre.[44] Her social ascension then readily translates into transnational ambitions, signified by her manager's earlier question, 'How would you like to try subduing London?'[45] Carrie's delocalized self-image complements the workings of the American System through a 'multipolar' representation of the world.[46] The rapid acceleration of the globalizing of national influence further worries Adams, as he links this modern trait to his increasing sense of global instability. Indeed, such acceleration could well usher in a postcolonialist phase, as Carrie's 'cyborg replication' could be read as performing the postcolonial idea of mimicry.[47]

Yet Adams's investment in techno-subjectivity has *already* entered into the realm of 'cyborg politics' (or biopolitics, an idea I will discuss below) which, sounding on a postcolonialist register,[48] interrogates the humanist conception of 'being' through the emergence of self-organizing systems. Adams and Carrie are hence unlikely bedfellows. In the technoscientific regime, the logic of representational power blurs the distinctions of absence and presence; such logic anticipates Jacques Derrida's interrogation of 'a politics of the virtual', of 'a certain delocalizing virtualization of the space of communication, discussion, publication, archivization'.[49] If the American System is a prototype of a virtual machine for creating and maintaining national influence, then its artistic counterpoint is Carrie's performance. And both shed light on how cultural imperialism *leverages* the principle of thermodynamic gradients (or differences) that informs the logic of global markets. The disparity figured in 'the surplus of power' that shapes the American System, according to Bové, reminds us that the economic *privilege* of

US citizens is created through the unequal conditions for so-called reciprocal trading agreements.[50]

For Adams, this surplus of power threatens to rewrite his precis for the future of American global empire into a consumerist fantasy of the new American woman as techno-subject. Indeed, Carrie's performance as this fantasy woman/cyborg pressures the self-similarity of national identity. As we have noted in previous chapters, for Judith Butler, '[P]erformativity is thus not a singular "act", for it is always a reiteration of a norm or a set of norms', while at the same time, 'reiterations are never simply replicas of the same'.[51] In *Sister Carrie*, the engineer (as scientific spokesperson), Bob Ames, tells Carrie that her face is a 'thing the world likes to see, because it's a natural expression of its longing'.[52] The surplus of power informs the crucial disparity in observation, the male gaze that translates the desire to see into the consumption of an image, what Cary Wolfe calls 'a consumerist schematization of visuality',[53] that animates and circulates through female bodies such as Carrie's whose performance reiterates the social norms of heterosexual, male, upper-class desire.[54]

In the virtual regime of techno-science, performance comes closest to realizing the longing for the 'real', yet can only be represented through the logic of self-organization, which transforms entropy in a closed system into a productive enterprise in an open system. As Livingston sarcastically remarks (directed, in particular, at theorists who misread thermodynamics as the latest capitalist rationale for social progress): 'Life, the tragic hero of the late nineteenth century, becomes at the dawn of a new millennium a surfing CEO with a cell phone'.[55] But 'life', overall, looks much different in Dreiser's novel. The attention shifts from Carrie's emergence as a transnational celebrity to the plight of the members of the underclass: 'They were of the class which simply floats and drifts, every wave of people washing up one, as breakers do driftwood on a stormy shore'.[56] This striking image of the inverse of Carrie's surfing in the thermodynamic economy – even, perhaps, suggesting that members of the underclass *are* the 'wave' upon which Carrie surfs – haunts more optimistic theoretical outcomes.

In Wharton's *The Custom of the Country*, Undine Spragg, who views marriage and divorce as complementary, rather than oppositional, gambits for social advancement, embodies the disruptive modern energies of Adams's dynamo.[57] Her performance brings together the nation's founding documents and the techniques of surfing over energy/economic flows through 'her inalienable right to "go around"'.[58]

The surplus of power that grants Undine social capital motivates the observations of Charles Bowen (the novel's de facto social reporter) about the modern transformation of marriage that shapes the new American woman. Yet his observations are connected to labour exploitation (that goes beyond US boundaries) in free-market sanctioned economic zones such as Africa.[59]

As an economic 'free agent' (who benefits from the *parallel* 'American system of easy divorces'),[60] Undine's transnational adventures remind us of the global systems proliferation of water management, national influence and artistic production. Her agency is enabled by her father's corporate deals that, energized by the local pull of the surplus of power, reclassify complex and troubled questions about ethics. As the narrator explains, Undine's father's 'private rule of conduct was as simple as his business morality was complicated'.[61] Ethics then becomes redefined as a performative category. As we previously noted, Wolfe comments that the functional differentiation of systems resists 'some untainted, unmediated space of subjunctive interiority . . . some unquestioned and unquestionable yardstick by which the socially and linguistically constructed code of ethics may be judged . . . the distinction between morality and merely ethical roles is lost'.[62] Undine's performance further destabilizes the idea of any intention *against* which moral judgement could be applied. Her desire does not reflect self-will, only some deferral of that self which 'speaks': 'I want what the others want'.[63] Here the deconstruction of identity, Derrida insists, 'prohibits us more than ever . . . from opposing presence to its representation, "real time" to "deferred time", effectivity to its simulacrum, the living to the non-living, in short, the living to the living-dead of its ghosts'.[64]

Therefore Wharton's novel threatens to reveal the ghost in the machine of the American System, disclosed by critical descriptions of the cultural production that the surplus of power enables. Bowen observes 'what unbounded material power had devised for the delusion of its leisure: a phantom "society", with all the rules, smirks, gestures of its model, but evoked out of promiscuity and incoherence while the other had been the product of continuity and choice'.[65] Undine's agency, her ability to 'go round', seems uncanny, phantom-like, as suggested by her posing for a portrait 'in a "dead white" . . . glamour'.[66] It is a performative reiteration of the blankness at the base of representational power that constitutes the modern development of technoscientific empire.

III. The American System and global debt

The American System's ongoing dependence on a surplus of power is counterpointed by rising global debt. If Adams, while opposed to socialism, is, like Hay, seeking non-militaristic strategies for national expansion, his historical speculations elide, as Livingston would point out, how 'a relentless aestheticization of the apparently kinder and gentler paradigms of chaos and complexity keeps the ecology of violence that it describes hidden in plain sight'.[67] One critical manifestation of such violence, Livingston says, is the sudden flight of capital from local economies. As James Heintz and Radhika Balakrishnan explain, African countries (shaped by the modern postcolonial legacy of labour/economic exploitation) in particular suffer from their social elites moving monies out of their countries, thus driving up local debt. These 'unrecorded flows' are enabled by 'international banks and financial institutions, usually based in the global North, whose cooperation was essential for capital flight to take place' and who, of course, profit from these illicit transactions.[68]

The patterns of acquisition that constitute thermodynamic self-organization, Livingston again would remind us, are modelled after the historical tactics of capitalist finance. At the end of Wharton's chronicle of the economic free agent, *The Custom of the Country*, Undine reaches her peak of social power through her continuous focus on self-investment. Yet one desire remains unrealized: becoming an ambassadress (and thus having more direct participation in the workings of global power) is impossible for her as a divorcée. Undine's performance hence reaches its limit: a pre-modern edict on marriage, that in part, establishes the legitimacy of the American System, also denies her 'the one part she was really made for'.[69]

For his part, Adams struggles with the effects of the modern regime of techno-science on his own view of the American System. In a 21 June 1905 letter to Charles Milnes Gaskell (written from Paris), Adams speaks of his observations on history (shaped by '[f]orty years' of life experience) as having become, in a sense, virtual: 'I suppose they were real; – but often I wonder if it is not my dream'.[70] Yet Bové draws attention to the performativity, the 'highly contrived' style, of his letter writing about the fate of national ambitions.[71] It is rather remarkable, I think, that Adams's white, male nostalgia for a unified state, seen as a performance that connects race, gender and class to empire, goes largely unremarked.[72] His forecast of apocalypse – that is, as Bové remarks, that 'eventually, the energy will dissipate and the system will crash'[73] – strategically uses thermodynamic imagery to defer the ethical debt compounded by the surplus of power that animates the American System. His insistence that time is running out, however, can only bring that final reckoning closer.

But here is another challenge that Adams anticipates: the erosion of boundaries between self and environment. To recount, in the *Education*, his construction of woman as Darwinian dynamo forecasts his anxieties about the 'permeable' body, a fantasy turned to nightmare in the image, in the 'Letter', of social systems that absorb human identity/agency.[74] This erosion is manifested (as discussed in Chapter 5) by the functional differentiation of systems-theoretical scenarios of economic advancement that have ecological consequences. We might say, then, that the materialization of systems brings such consequences into sharper relief. In Alaimo's reading of Karen Barad, matter itself is active, not passive. As a result, there is not a subject in possession of agency and set apart from the environment; rather there is an 'agential' material world in which 'things, as such, do not precede their intra-actions'.[75] Similarly, Alaimo proposes that Beck's 'risk society' implies that it is difficult for us to calculate the absolute probability of harm. Collectively Barad and Beck invoke what Alaimo calls trans-corporeality: a condition that denotes the openness of human bodies and non-human natures to one another, so that when we pollute the soil, water and/or air, we thus pollute our own bodies. In particular, Barad's idea of performativity (which as we saw in chapter 4 is informed by techno-science and pushes further Butler's model) is crucial for understanding not just how the 'new woman' was already the prevailing model for national techno-subjectivity, but the posthuman turn from the secure liberal humanist body to a trans-corporeal body.[76] We might surmise that geopolitical control, focused on the policing and expansion of national borders, thus applies to the biopolitics of bodies.

From a biopolitical perspective on 'establishing spheres of influence', there is increased anxiety about national identity – for instance, when remapped on a global scale – not being self-identical, registering as the ghost in the machine (or a manifestation of Adams's 'truism' about force surpassing intelligence). Furthermore, the American System's accelerated model of economic growth boosts consumption, remapping and re-territorializing postcolonial geographies into sites of capital flow and what Michael J. Watts describes as 'oil frontiers' that involve 'relations between the deep infrastructures of the oil world – pipelines, rigs, flowstations, tankers, financiers, engineering firms, security forces, and so on – and to the regimes of life and death in the postcolonial South and the advanced capitalist North'.[77] To look more closely at these frontiers does not paint a promising picture of the future. One pertinent example is Nigeria: its 'spectacular petro-capitalism combines the most brutal forms of capitalist dispossession and racialized accumulation with the ecological wreckage of a modern high-tech global oil and gas industry'.[78]

To the ethical issues surrounding 'capitalist dispossession' and 'racialized accumulation' can be added the ongoing threat of the parasitical attachment of (business) morality. Glimpsed, for instance, in *The Custom of the Country*, the parasiting of systems by a moral injunction for Luhmann would set the limits on an ethics of biopower. That is, Wolfe elaborates, Luhmann would say that the 'self-reference of the law's autopoiesis protects the communicative actions that take place in the legal domain from being steered or recoded by the moral distinction of '"right vs wrong"'.[79] Wolfe goes on to say that 'a permanent disjuncture between ethics and law' joins Luhmann with Derrida: 'For both, the law is radically aporetic or paradoxical in its self-reference and immanence; the difference between legal and illegal is (paradoxically) itself instantiated by the legal, hence the tautological self-reference, "legal is legal"'.[80] Thus the 'disjuncture between ethics and law' is reproduced within discourse about the American System and the future of foreign policy. To take just one example, Bové quotes Robert Cooper, a highly influential British foreign policy strategist:

> The challenge to the postmodern world is to get used to the idea of double standards. Among ourselves, we operate on the basis of laws and open cooperative security. But when dealing with more old-fashioned kinds of states outside the postmodern continent of Europe, we need to revert to the rougher methods of an earlier era – force, pre-emptive attack, deception, whatever is necessary to deal with those who still live in the nineteenth-century world of every state for itself. Among ourselves, we keep the law but when we are operating in the jungle, we must also use the laws of the jungle.[81]

Further destabilizing the future of global geopolitics, we can see the escalation of 'risk production' the oil frontier portends;[82] just as the petro-frontier globally extends the conflicted social landscape of the naturalistic, post-frontier American West in the early twentieth century, a global framing of biopolitics envisions the Pure Water Move (in *The Custom of the Country*) as the precursor to the looming threat of global ecosystemic collapse through oceanic pollution and overfishing.[83] Viewed through the lens of posthuman ethics, these complex risk-scapes suggest, as Jane Bennett and Barad would surely agree, that we are no longer exclusively privileged agents. Take, for instance, eating: a situation where, Bennett writes,

> human and nonhuman bodies recorporealize in response to each other; both exercise formative power and both offer themselves as matter to be acted on. Eating appears as a series of mutual transformations in which the border between inside and outside becomes blurry: my meal both is and is not mine: you both are and are not what you eat.[84]

Yet as we will see in more detail, Bennett's description does not pay adequate attention to a pressing issue: how humans manipulate other agents and environments through corporate agribusiness and factory farming, for example. The field, even with an account of redistributed agencies, remains far from level. Attempts, furthermore, to make matter matter (what has been termed New Materialism) tend to resemble, as Heise argues, a return to the Deep Ecology thinking of the 1970s and 1980s which idealistically claimed that all life forms were equal, only to run up against insurmountable problems in defining the criteria for valuing life.[85] These problems, summed up by Wolfe, are the following: should rare or endangered species be prioritized over humans? Do dangerous viruses have a right to flourish unchecked? And, of course, demographics: how to account ethically for the fact that in a zero-sum global struggle for survival, so-called First World nations would always have an unqualified advantage.[86] Hence Bennett's closing creed, 'I believe it is wrong to deny vitality to nonhuman bodies, forces, and forms',[87] does not square an ethical stance with how 'vitality' is to be recognized. For her part, Barad, while asserting that 'learning how to intra-act responsibly as part of the world means understanding that "we" are not the only active beings – though this is never justification for deflecting our responsibility onto others',[88] leaves unexplored the relationship between agency and responsibility, much less the relationship, as Heise would remind us, between agency and manipulation. Thus Alaimo gets to the heart of the matter by putting the idea of agency (as an ethical given) into question: '[O]ne of the central problematics of trans-corporeality is contending with dangerous, often imperceptible material agencies. The sense of being permeable to harmful substances may provoke denial, delusions of transcendence, or the desire for a magical fix . . . but it may also foster a posthuman environmentalism of co-constituted creatures, entangled knowledges, and precautionary practices'.[89] What Alaimo is getting at here is a sense of our vulnerability: a new way of modelling the posthuman that Wolfe will now take up.

IV. Biopolitics and posthuman life: the call of Jack London

For Wolfe (guided by Derrida), what establishes the openness to cross-species affiliations is our sense of shared vulnerability, on multiple levels, with non-human animals. Deriving from Jeremy Bentham, what does 'trace the insuperable line' that establishes whether beings (humans or non-human animals) are within the parameters of ethical consideration is the question, 'can they suffer?'[90] By asking 'what of the vulnerability felt on the basis of this inability?',[91] Derrida makes it clear that the issue of ethics stems not from being able to act (or signal an intention to act), but from precisely the opposite condition. According to Derrida, language, however, gets in the way of our articulating this vulnerability, which we share with animals. With Derrida, Wolfe elaborates that we have two types of

> passivity and vulnerability. The first type (physical vulnerability, embodiment, and eventually mortality) is paradoxically made unavailable, *inappropriable*, to us by the very thing that makes it available – namely a second type of 'passivity' or 'not being able,' which is the finitude we experience in our subjection to a radically ahuman technicity or mechanicity of language, a technicity that has profound consequences, of course, for what we too hastily think of as 'our' concepts, which are therefore in an important sense not 'ours' at all.[92]

And this second 'finitude' is what Derrida will use systematically to undermine claims for a conscious sense of self (variously expressed as consciousness, reflection, signification, concept-making, second-order understanding, metacognition and so on) which is said to be the attribute *exclusively* of humans.

Because, as Wolfe reminds us, our concepts are not ours at all, we are compelled to revisit the controversy raised by Jacques Lacan. Wolfe explains that Lacan's 'The subversion of the subject and the dialectic of desire in the Freudian unconscious' (1960) 'juridically separate[s] the human from the animal as that being, alone among the living, who can *lie by telling the truth*. The animal, in Lacan's terms, can pretend, but not *pretend to pretend* – only the human, as "subject of the signifier," can do that'.[93] On that basis, Derrida advises, it is argued that an animal can only react, not respond; in other words, an animal is unable 'to *pretend*, to *lie*, to *cover its tracks* or *erase* its own tracks'.[94] To put it schematically, Derrida counters Lacan by proposing that the 'trace' be 'substituted' for the signifier, so, as Derrida says, 'the structure of the trace presupposes that *to trace* amounts to *erasing a trace* as much as to imprinting it',[95] and as he contends, 'these possibilities or necessities, without which there would be no language, *are themselves not only human*'.[96] According to Wolfe, then, we share with animals this second 'passivity and subjection', 'a more radical "not being able" that makes our very conceptual life possible', that strikes at the heart of the question of who we are: '[W]e are always radically other, already in- or ahuman in our very being . . . in our subjection to and constitution in the materiality and technicity of a language that is always on the scene before we are, as a precondition of our own subjectivity'.[97] With that, the idea of our being able to express our intentions, what we mean, through language (or, similarly, that language can express our intentions), does not offer any pretence of using intention in language to support the supposed cognitive differences between humans and animals: what Wolfe calls the fantasy that second-order meanings, constructed as 'the distinction between superficially "meaning it" and *really* "meaning it"',[98] are exclusively human. As Wolfe later puts it, what is absolutely critical here is 'the concept of "the human" that the human "falsely gives to itself" to then enable its recognition – from a safe ontological distance – as it were – of the nonhuman other in a gesture of self-flattering "benevolence" wholly characteristic of liberal humanism'.[99]

If we are unaware of the stakes for which such 'recognition . . . of the nonhuman other' (that would lead to legal/ethical consideration) transpires, we may revisit Adams. As he writes, in an 11 August 1905 letter to Margaret Chandler: 'America has nearly a hundred million people running at least five-and-twenty million horse-power equivalent to the power of the whole animal world since Eve'.[100] Reading this letter through the *Education*, as Pynchon would suggest (in *Against the Day*), Adams's dynamo stages the same 'reduction of choices' as 'herds of cattle' would experience, 'until the final turn through the final gate that led to the killing floor' of the Chicago slaughterhouse.[101] Glossing this idea of the slaughterhouse in *Before the Law*, Wolfe explains that the genocidal practices in Nazi concentration camps 'derived from production models originally developed by Henry Ford (a notorious anti-Semite)' that were inspired by the 'mechanized *disassembly* line' of the Chicago slaughterhouse.[102] Indeed, factory farming is, at base, a further exercise of 'the technological manipulation of life' that 'does not abide by the logic of genocide', that is, species extinction, but by (in Derrida's words) the logic of a 'virtually interminable survival in conditions that previous generations would have judged monstrous'.[103]

It is here that our perspective crucially shifts. The Holocaust is, to be certain, a critical entry on a list of racially inflected effects of biopolitics; indeed, Giorgio Agamben, as Wolfe points out, sees 'genocide as the underlying paradigm and constitutive tendency of modernity'.[104] Yet following Wolfe, to talk about race and biopolitics, as it shapes, for instance, capital flight and oil frontiers, 'you can't talk about race without talking about species, simply because both categories – as history well shows – are so notoriously pliable and unstable, constantly bleeding into and out of each other'.[105] We thus re-focus on the 'animal question', for Derrida, a conflict 'between what calls *itself* man and what *he* calls the animal'.[106] And a central aspect of that question is what we make of the supposed differences between humans and animals. Wolfe's earlier commentary on how language always/already is there 'before us' radically troubles how we distinguish between what is thought of as human 'response' and animal 'reaction'. If language can no longer indicate our intention (if we can no longer be said to be in control of language, language no longer proves what we mean to say or do), then as Wolfe's reading of Derrida contends, we no longer have exclusive access to the power of signification, of making meaning – in short, to respond. To put it slightly differently, we must ask, as Derrida does: can there be a meaningful response (expressed in a language that, as we know, is always/already before us) that somehow does not carry a trace of automaticity, of a reaction?

Such a loss of exclusivity guides Wolfe's study of the legal dimension of the animal question. He traces a sacrificial economy that guides the construction of law on the basis of exclusion:

> [N]amely the violence of sacrifice for which the distinction between human and animal has historically been bedrock, providing for the law the 'foundation' for its exclusions that the law cannot provide

> for itself. As Derrida, Agamben, and others have reminded us, those who fall outside of the frame, because they are marked by differences of race, or species, or gender, or religion, or nationality, are always threatened with 'a non-criminal putting to death'.[107]

In a rather precise way, Wolfe examines how the issue of (human) response versus (animal) reaction registers in biopolitical terms, starting by defining what life can be sacrificed that constructs the notion of sovereignty (following Agamben) and then moving to Derrida's critique of how 'the designation "the animal" . . . effaces the vast diversity of nonhuman life under a single definite article'.[108] As Wolfe posits, in his reading of Derrida's recently translated lectures, Derrida relates the question of 'what calls *itself* man' to the preservation of the traditional opposition between 'the side of the "who" and the side of the "what"'.[109] What Wolfe crucially brings out is how 'contemporary research in philosophy of mind, cognitive ethology, and animal communication' gives weight to the contention 'that what we think of as the capacity to "respond" is the product of a complex, dynamic, mutually imbricated relationship between the "who" and the "what" – a fact that applies to both humans and (at least some) nonhuman animals'.[110]

At this point, it should be noted that Derrida's position (which is to reject 'some homogenous continuity between what calls *itself* man and what *he* calls the animal') has as its objective 'not to efface the limit' between different forms of life, 'but to multiply its figures, to complicate, thicken, delinearize, fold, and divide the line precisely by making it increase and multiply', what we just saw in the multiplicity of the 'animal'.[111] To continue, the second vulnerability (our subjection to language) does not then mark *the* limit of shared relations between human and animal, *the* difference between human and animal. It does not support what Wolfe describes as 'one of the more ingenious humanist responses to the "delinearizing" of the line between forms of life', that is, 'the response that the distinctly human is constituted precisely by a radical not being-able that is barred to other creatures; in [Martin] Heidegger, the "letting be of Being" (but also in a different register, the existential Being-toward-death); in Agamben *Dasein* as that being who has "awakened *from* its own captivation *to* its own captivation"'.[112] This is because Derrida, as Wolfe reminds us, insists that

> this essential '(not) being-able' is not the prerogative or even the place, you might say, of either the subject or indeed of 'the human,' since

> it is to be located elsewhere, in an alterity that is not just radically extrasubjective, and not just radically ahuman, but also, in fact, radically *inorganic*: namely, in the prosthetic relation to the externality and technicity of trace, archive, symbolic system, or semiotic code, however rudimentary. The subject, in other words, comes into being only by virtue of what it is not – what it is *radically* and *absolutely* not – rather than by virtue of something it *chooses* not to be (even in the mode of misrecognition, as Lacan et al. would have it), something over which – either through intentionality or unconscious disavowal – it has appropriative power.[113]

Or as Wolfe puts it, because 'the other is just as constitutively other to itself as I am to myself', '*any* "who" becomes one only by virtue of also being, prosthetically, a "what"'.[114] And when it comes to issues that pertain to deep ecology (from a biopolitical perspective), voiced as 'questions of value', Wolfe proposes that these questions

> necessarily depend on a 'to whom it matters,' but that 'to whom' need not be – indeed as we have already seen, *cannot* only be – human, either in the sense of excluding by definition nonhuman animals, *or* in the sense of a 'human' who is not always already radically other to itself, prosthetically constituted by the ahuman and indeed inorganic. If the capacity 'to respond,' to be a 'to whom,' is not given but emerges, is brought forth, out of a complex and unfolded relation to the 'what,' to its outside (whether in the form of the environment, the other, the archive, the tool, or the 'instinctive' program of behavior), then the addressee of value – and indeed of immunitary protection – is permanently open to the possibility of 'whoever it might be'.[115]

To take this 'possibility' as a legal question, as Wolfe advises (and has been addressed elsewhere in this book), Luhmann would say that the legal system is protected from being influenced by moral distinctions.[116] As Wolfe states earlier, this 'permanent disjuncture between ethics and law' connects Luhmann and Derrida.[117] And yet, following Luhmann (and Derrida), Wolfe counsels that law and justice are not 'in a relationship of opposition, exactly', because of what is 'after' the law, its 'conditional' or *performative* character based on 'the fact that the inescapable self-referential closure of law is precisely, in its contingency, what opens it to the future and the outside, the as yet unaddressed subjects – who knows how many? – of justice'.[118] So while we might say the case is *not* closed when it comes to matters of justice (the open question of 'how many?'), there is a second, *sacrificial*, limit that relates (in biopolitical terms) to the first. We have already seen this limit as it has traced the connections between the Chicago slaughterhouse, Nazi concentration camps and factory farming. Now in light of the question of the performativity of the law (or as, Luhmann would say, how the law does what it does), such a limit becomes crucial for understanding the construction of sovereignty.

To begin, the larger idea of the political (which is, it should be noted, not autonomous, being itself a system self-referentially closed) sustains 'sovereignty in several different registers', Wolfe remarks, 'not just of the nation-state but of the family . . . the "proper" to man, the *oikos* of the ecological'.[119] He goes on to remark that 'the concept of "the human" that the human "falsely gives to itself"', that we have previously discussed, serves schematically as the basis for 'the autonomous subject . . . whose very paradigm is the "head" of State'. Wolfe speculates, with Derrida, that the basis for sovereignty depends not upon sacrifice 'in some abstract sense but "carnivorous sacrifice" as testament to its autonomy'.[120]

If we *do* take this speculation seriously (and I, for one, think we should), then the notion of what life can be sacrificed changes. To quote Derrida: '[L]ife has absolute value only if it is worth *more than* life . . . It is sacred, holy, infinitely respectable only in the name of what is worth more than it and what is not restricted to the naturalness of the bio-zoological (sacrificeable)'.[121] Derrida follows this logic to its endpoint: the 'respect of life' ('Thou shalt not kill') can apply 'only in so far as it bears witness, in some manner, to the infinite transcendence of that which is worth more than it (divinity, the sacrosanctness of the law)'.[122] Thus Derrida pressures Luhmann's argument that the law is protected against moral infection by functional differentiation, because, as it turns out, religion and politics are constructed through the same logic of sovereignty; as Wolfe comments, 'they inflect and "infect" each other'.[123]

From this viewpoint, globalization does not set religion and politics (the law) against each other. Indeed, as Derrida would pessimistically assert, there are no limits, biopolitically speaking, to a global campaign of capitalist expansion and acquisition. What therefore seems crucial to the economic and ecological turmoil generated by capital flight and oil frontiers (in particular as they relate to the American System) is, Wolfe explains, 'the complex psychodynamics of sacrifice and eating flesh as a sign of the subject's autonomy, security . . . a key export for the success of capitalist globalization, one borne on the backs of billions of dead animals'.[124] To find support for Wolfe's argument, we simply need to look at the case of China. Now a major player on the world stage, the country has, following the capitalist trend of economic modernization, radically altered its food consumption patterns. Historically known for its 'predilection' for vegetarianism, it has changed to meat eating, despite 'the extreme inefficiency of meat as a food source' (due to its population size and lack of farm-land); consequently, 'China's meat mania is implicated in everything from deforestation in Brazil to food-price inflation in Africa'.[125]

This current state of biopolitical crisis warrants a rethinking of London's honest embrace of how animals are, as the title of Wolfe's latest book states, *Before the Law*. By this, Wolfe means, first, 'the sense of that which is ontologically and/or logically antecedent to the law, which exists prior to the moment when the law, in all its contingency and immanence, exacts its originary violence',[126] that is, establishes the distinction (or a 'frame') between who is granted legal protection and who is not. Second, Wolfe means 'the sense of standing before the judgement of a law that is inscrutable not just because it establishes by fiat who falls inside and outside the frame, but also because it disavows its own contingency through violence; namely, the violence of sacrifice for which the distinction between human and animal has historically been bedrock'.[127]

In London's 'The Other Animals', *The Call of the Wild* and *White Fang*, he relentlessly draws attention to the animal question, avoiding the humanist 'deflection' or even 'disavowal'[128] inherent in Adams, Wharton and Dreiser's treatment of biopower – even as London's reactionary politics (of nation, race and gender) pose, to be sure, a challenge to the rethinking of species distinctions. As we have tracked the animal question, the image of 'drawing the line' (that determines whether one deserves ethical consideration) emerges time and again. To review briefly: such an image surfaces in Bentham's proposal that the capacity to suffer is what would 'trace the insuperable line', in the 'materiality' of Derrida's trace and especially in Derrida's efforts to 'delinearize' and 'multiply' the line between 'what calls *itself* man and what *he* calls the animal'. London is already thinking about this issue; as early as 1899, his letter to Cloudesley Johns stages the animal question and expresses a biopolitical linkage between species and race: 'If no race has the moral right to survive in the destruction of lesser races, where, in the whole animal kingdom are you going to draw the line?' London repeats the image, in more detail, in the same letter:

> Do you know that the physiologists say that the difference between the highest forms of man and the lowest forms of man is greater than the difference between the lowest forms of man and the highest forms of the rest of the vertrebates [*sic*]? This being so (and it is so), where are you to draw the line? If the lowest forms of man are to be preserved, no matter what the moral logic involved, then, by that same logic, you must not exclude the higher forms of the rest of the vertrebates [*sic*].[129]

From this viewpoint, *White Fang* crystallizes and dramatizes London's thinking; the wolf protagonist, after attacking one of Weedon Scott's dogs, goes after Matt, Scott's dog-masher, prompting Scott to say: 'It's all right about the dogs, but we must draw the line somewhere'.[130] As we are reminded, one way to formulate the 'line' is to conceive it as marking the difference between reaction and response (as Lacan famously tried to do). By focusing, in both novels, on the trainability of dogs, London shifts the terms of debate, for animal trainers tend to speak of a response of an animal rather than its reaction. For example, referring to 'the mythology of Buck' in *The Call of the Wild*, Vicki Hearne posits that a 'dog or horse trainer's special interest in "wild and uncontrollable" animals assumes that the degree of violence in a dog's resistance to authority may indicate the depth and power of response such a dog might give when the gap between command and obedience is closed'.[131] We recall here Wolfe's citing of recent research that further questions the notion of response as an exclusively human capability; growing evidence suggests it is instead 'the product of a complex, dynamic, mutually imbricated relationship between the "who" and the "what" – a fact that applies to both humans and (at least some) nonhuman animals'.[132]

In *White Fang*, London looks further at the ethical implications of regarding dogs as companion species. Although White Fang 'had no abstract ideas about justice and fair play', he retains 'a sense of equity'.[133] This 'sense of equity' is shaped by his decision to sit by the fire of Gray Beaver, his first master: 'At last he lay at the master's feet, into whose possession he now surrendered himself, voluntarily, body and soul. Of his own choice he came in to sit by man's fire and to be ruled by him'.[134] But how London, through White Fang, perceives racial differences determines the ranking of those men who rule; as compared with Native Americans, white men are 'a race of superior gods'. Of course, White Fang 'does not reason it out'; it is a 'feeling'.[135] There is a related difference in his feelings of affection. White Fang's strongest emotional attachment is to his last owner, the white, upper-class mining engineer, Weedon Scott, 'the love-master'.[136] The narrative intends to depict love as a natural trait; Charles Darwin writes that the 'lower animals' exhibit the 'same principle of pleasure' as humans do.[137] But at the same time, it is an unnatural trait, for Mark Seltzer states: 'Learning to love pain and the god-like hand of his master, White Fang learns to love at once the pleasure of unnatural acts (acts contrary to every "mandate of his instinct") and the pain of turning from "the natural" to the "cultural"'.[138]

The civilizing of White Fang blurs the distinctions between reaction and response in two crucial scenes. When Weedon Scott is thrown from his horse and breaks his leg, White Fang runs to the house and is able to communicate, through barking, what has happened. Indeed, it seems clear that this is more than just reaction: even once domesticated, as a wolf, White Fang has difficulty barking. As the narrator states: 'For the second and last time in his life he had barked and made himself understood'.[139] This moment of communication shows us how a wolf can use language to make '*him*self understood', a process of signification that revisits Derrida's view of the animal question as a conflict about who/what is able to declare him/it*self*. Likewise, in the final scene, the notion of love is set up as something responded rather than reacted to. A brutal convict, Jim Hall, escapes from prison and seeks revenge against Judge Scott, who had presided at his trial. When Hall breaks into the Scott household, the wolf waits until Hall starts up the stairs that 'led to the love-master and his dearest possessions' before attacking.[140] In London's description of White Fang here, we see that he acts equally out of instinctual reaction and as a response to his love for Weedon Scott.

Furthermore, we get a wolf's view of how biopolitics determines what counts as life that may be sacrificed: 'In the end he worked out the complete law. Between him and all domestic animals there must be no hostilities . . . But the other animals – the squirrels, and quail, and cottontails, were creatures of the Wild who had never yielded allegiance to man'.[141] In London's narrative, we observe the dual vulnerabilities that constitute life in a biopolitical domain: the recognition of the first, physical vulnerability, is enacted through the act of 'yielding allegiance' (as White Fang does) that leads to the second, the wolf's acceptance of the performativity or contingency of the law, secured through the act of sacrifice. In this light, the biopolitical trajectory of the novel requires White Fang to offer to sacrifice *himself* in the act of defending Judge and Weedon Scott, patriarchs of a family that cuts across species lines while enforcing distinctions between species (as what can be killed without retribution). And in the outcome, White Fang's sacrifice become more symbolic (for he does not die), as the Judge names him (through language) as 'Blessed Wolf'; in turn, White Fang becomes a father himself, thus reproducing the patriarchal values of sovereignty that he has internalized.[142] It is, moreover, a sovereignty that blends religion and law, as performed through the Judge's blessing of the wolf and upholds the wolf's masculine subjectivity within a cross-species family.

Note, indeed, the biopolitical framing of the two novels; complementing the supernatural outcome/destiny of Buck becoming 'Ghost Dog', the opening of *White Fang* states:

> [L]ife is movement; and the Wild aims to destroy movement. It freezes the water to prevent it running to the sea; it drives the sap out of the trees till they are frozen to their mighty hearts; and most ferociously and terribly of all does the Wild harry and crush into submission man – man who is the most restless of life, ever in revolt against the dictum that all movement must in the end come to the cessation of movement.[143]

In his interpretation, Seltzer comments that 'the mastery of men – the *un*natural mastery that revolts against the natural law of "the Wild" that freezes motion – consists of technologies of flow and continuous motion'.[144] Eugene Thacker says: 'The contradiction is clear: life is that which renders intelligible the living, but which in itself cannot be thought, has no existence, is not itself living'.[145]

In the ending of *White Fang*, as the wolf's life dramatically hangs in the balance (suspended between life and death), the dream of the squirrel becoming an electric car reflects the fantasy of an animal outsider/other – by virtue of its not having chosen 'allegiance to man' – that mediates the 'contradiction' by reinforcing, as Wolfe points out, Heidegger's notion of the 'animal-machine' that is 'neither living nor dead' and can only react, not respond – and certainly not choose.[146] That this vision of White Fang's is rather fleeting suggests the answer to the animal question that London will give in 'The Other Animals'. London anticipates Wolfe's argument about the 'recognition' of the animal other. For Wolfe, in a liberal humanist register (guided by

Heidegger), the human 'is capable of a more or less transparent relationship to its own nature, a relationship that it *then* expresses in and through language and may *then* extend benevolently – or not – to the nonhuman other'.[147] And here is London:

> You must not deny your relatives, the other animals. Their history is your history, and if you kick them to the bottom of the abyss you go yourself. By them you stand or fall. What you repudiate in them you repudiate in yourself – a pretty spectacle, truly, of an exalted animal striving to disown the stuff of life out of which it is made, striving by use of the very reason that was developed by evolution to deny the process of evolution that developed it.[148]

Yet what of the pragmatics of London's biopolitical imperative? To take the animal question seriously, of course, requires us to think about the ethical treatment of animals, which must include, Wolfe reminds us, some form of legal protection. To address the fundamental problem of biopolitics: as Wolfe explains, there is a (false) dichotomy stated as '*either* unconditionally embrace all forms of life as subjects of immunitary protection, *or* suffer the autoimmune consequences that follow';[149] this is what we previously saw in the doctrine of sovereignty stating that 'life has absolute value only if it is worth more than life'. To value all life equally is an act of bad faith, because, as we have discussed, it flattens species differences completely and crucially elides the sense of identity for any animal (and wrongly reduces the diversity of animal life to a single, amorphous designator). As Wolfe concisely puts it, 'we *must* choose, and by definition, we *cannot* choose everyone and everything at once'.[150] But as Wolfe points out, because we must take into account what comes after as well as before the law, 'this very act of immunitary selection and protection on the basis of the capacity to "respond" . . . can never be juridical, however, because it is always already traced with the automaticity and mechanicity of a reaction'. Wolfe then states the act can be most accurately described as 'a "line", to use Derrida's formulation, that is always already "multiple" and nonlinear, always folded and in motion, always under erasure'. And so the conclusion that Wolfe draws, following Derrida, is about a reaction that will turn out to be always/already inadequate to meet the pressing demands of the biopolitical moment: '[*I*]*n* the future, we *will have been wrong*. Our "determinate" act of justice now will have been shown to be *too* determinate, revealed to have left someone or something out'.[151]

EPILOGUE

We should take a moment to reflect on the central goal of the project: to resituate the prose narratives of Dreiser, Wharton, London and Adams within a crucial moment when evolutionary thought and a shifting picture of the world destabilized social categories of race, class, gender and citizenship. To be sure, Wolfe's argument about biopolitics is precisely a current (and rather pressing) version of destabilization reflected by naturalist texts. Wolfe's particular shock to the system resonates with Adams's closed system of paranoia: for Adams's image of society as an organism (that appears constituted through the automaticity and mechanicity of a reaction) increases a sense of vulnerability (Derrida's concept of 'not being able'): '[T]he social organism, in the recent views of history, is the cause, creator, and end of the Man [*sic*], who exists as only a passing representative of it, without rights or functions except what it imposes'.[1]

In Dreiser's *Sister Carrie*, a biopolitical logic shapes the end of Hurstwood's life. His seemingly willing surrender to death stages Foucault's pronouncement that 'the very fact that you let more die will allow you to live more'.[2] Here we may revisit Gavin Jones's interpretation of Dreiser's narrative: 'Nobody would be turning on the gas in an official capacity; this would not be genocide, technically speaking. More conveniently still, in this world of "moderate" organization the conditions of poverty would be allowed to take care of themselves'.[3] But increase, if you will, the level of 'organization', and you end up with Lily's fate in *The House of Mirth*. For Lily is trapped in a destructive economic system that is cast as distinctly inhuman, because (in accordance with Derrida) of the use of language acts to carry out its operations based on class-, race- and gender-based, yet contingent, social codes. In this light, Lily's destruction of the letters from Selden to Bertha Dorset (that might save Lily) presages the detachment (signified by his own inability, his not being able, to save Lily) of Selden, who at the end can only imagine a resolution for Lily's mysterious death: 'the word which made all clear'.[4] Through biopolitics, we can then trace a connection through the naturalist interrogation of agency, further pressured by a modern sentiment of 'convenience', where social problems 'take care of themselves', to Wolfe's destabilization of the 'consumerist "choice"' that fuels the ecologically destructive impulses of hyper-capitalism: '[W]e *will have been wrong*'.[5]

While we have seen in Wharton and Adams scenarios of the historical erasure of races, such as Native Americans, only in London do we observe such violence crossing species lines. London goes further, however, in addressing the legal sanctioning of violence. In *White Fang*, while London expresses some sympathy for Jim Hall, 'ill-made in the making' and 'innocent of the crime for which he was sentenced', it is implied that Hall, 'a human beast', is a threat to civilized society who must be violently controlled.[6] The sense of animal 'equity' (and the subsequent endorsement of upper-class values of property) is maintained over that of 'justice' for the wrongly convicted Hall, who is marked, or 'othered', as lower class. Yet as Wolfe reminds us, there is no legal resting point for the conflict London invokes between equity and justice, because the contingency of the law unfolds 'the inseparable possibilities of both promise and threat'.[7] Certainly within London's own drawing of the 'line', there is violence that hints at the eventual cost of sovereignty. To put it thusly, how can such distinctions (centred on the determination of who can be killed without retribution) necessary for sovereignty, in the biopolitical register, be preserved except by the law, whose operations are far from discrete? The *exposure* of this question pushes us in the direction of Wolfe's argument that 'the political vocabulary of sovereignty we have inherited' obscures 'new forces and actors' on the biopolitical stage: what Wolfe calls the emergence of 'a newly expanded community of the living' that positions 'all' of us as 'potentially animals before the law'.[8] Thus we need to take up the challenge that London puts forth: to see the dramatization of the animal question on a biopolitical stage which destabilizes the who that is observing and the what that is being observed.

NOTES

Introduction

1 My study of popular science in the US is interested in how consensus-building about scientific theories is a dynamic, unstable and unpredictable process, focusing on debates about the meanings of scientific models of biological development. Previous studies of popular science tend to look at scientific knowledge as already having been codified. Marcel C. LaFollette, *Making Science Our Own: Public Images of Science, 1910–1955* (Chicago and London: University of Chicago Press, 1990) concentrates on how a shared culture emerges between scientists and the public at large, but does not consider how distinctions between the scientific and social realms had become destabilized by the early twentieth century. While the author describes the role science plays in national affairs, she suggests that early-twentieth-century science 'was less entwined with national interests' (p. 16). On this matter, Ronald C. Tobey, *The American Ideology of National Science, 1919–1930* (Pittsburgh, PA: University of Pittsburgh Press, 1971) is still more dismissive, claiming that 'in the seventeen years preceding America's entry into the First World War, popular interest in science and professional popularization remained at a low level' (p. 3). My contention (which will be explored in detail in the first chapter) is that science already is firmly connected with national interests and had gained a significant degree of public interest by the early twentieth century.

My book participates in the conceptual rethinking of the study of popular science. In the introduction to 'Focus: historicizing "popular science"', *Isis*, 100/2 (June 2009), 310–18, Jonathan. R. Topham proposes that we follow the counsel of James A. Secord, whose keynote address, 'Knowledge in transit', appears in revised form in an earlier volume of the journal: *Isis*, 95/4 (December 2004), 654–72. Secord's plenary lecture was delivered on 6 August 2004 at the fifth joint meeting of the British Society for the History of Science, the Canadian Society for the History and

Philosophy of Science/La Société Canadienne d'Histoire et de Philosophie des Sciences, and the History of Science Society. He remarks that the 'positivist' belief that '[s]cientific knowledge spread because it was true; any failure of diffusion could be explained by resistance due to false beliefs and irrational commitments . . . would now have few supporters' (655). We are thus compelled to 'realize the centrality of knowledge in circulation – of science as a form of communication'. Although Secord does provide a view of science in action, in the making, that can traverse disciplinary boundaries, Bernadette Bensaude-Vincent thinks his proposal does not go far enough. In her response, 'A historical perspective on science and its "Others"', *Isis*, 100/2 (June 2009), 259–68, she points out that 'science is more than the production and communication of factual data' (361), calling for a further examination of how the categories of science and the public construct one another.

2 Alan Nourie and Barbara Nourie (eds), *N. W. Ayer & Son's American Newspaper Annual and Directory* (Philadelphia: N. W. Ayer & Son, 1895–1905); *American Mass-Market Magazines* (New York and Westport, CT: Greenwood Press, 1990), pp. 385–7.

3 The debate about social Darwinism in the US is treated by Robert C. Bannister, *Social Darwinism: Science and Myth in Anglo-American Social Thought* (Philadelphia: Temple University Press, 1979) and Richard Hofstadter, *Social Darwinism in American Thought* (Boston: Beacon Press, 1955). Hofstadter believes that social Darwinism permeated US culture and provided a rationale for conservative business practices (cut-throat capitalism), political agendas (imperialism and colonialism) and social policies (slavery and racial discrimination). Bannister, conversely, insists that social Darwinism was a much less influential idea in popular thought, arguing that it was largely a derogatory label used to discredit intellectual rivals.

4 Refer to Priscilla Wald, *Constituting Americans: Cultural Anxiety and Narrative Form* (Durham, NC: Duke University Press, 1995); Alan Trachtenberg, *Shades of Hiawatha: Staging Indians, Making Americans, 1880–1930* (New York: Hill and Wang, 2004); Walter Benn Michaels, *Our America: Nativism, Modernism, and Pluralism* (Durham, NC: Duke University Press, 1997); Paul A. Bové, 'Henry Adams and the American System', in William Merrill Decker and Earl N. Harbert (eds), *Henry Adams and the Need to Know* (Boston: Massachusetts Historical Society, 2005), pp. 1–17.

5 Informative discussions of these writers are featured in Jay Martin, *Harvests of Change: American Literature, 1865–1914* (Englewood Cliffs, NJ: Prentice Hall, 1967); Donald Pizer (ed.), *The Cambridge Companion to*

American Realism and Naturalism: Howells to London (Cambridge: Cambridge University Press, 1995); Ronald E. Martin, *American Literature and the Universe of Force* (Durham, NC: Duke University Press, 1981). My project takes an unexplored path, with regards to recent critical studies of Dreiser, Wharton, London and Adams. Jennifer L. Fleissner, *Women, Compulsion, Modernity: The Moment of American Naturalism* (Chicago: University of Chicago Press, 2004) looks at the psychological construction of gender that informs these writers' works without completely extending this reading to the interplay between race and citizenship categories that shape identity. Walter Benn Michaels, *The Gold Standard and the Logic of Naturalism: American Literature at the Turn of the Century* (Berkeley: University of California Press, 1987); Mark Seltzer, *Bodies and Machines* (New York and London: Routledge, 1992); Bert Bender, *The Descent of Love: Darwin and the Theory of Sexual Selection in American Fiction, 1871–1926* (Philadelphia: University of Pennsylvania Press, 1996) all examine the social contexts for these writers, but tend to focus on their reception of scientific concepts, without considering how the conditions for understanding science in the early twentieth century were shifting and unstable. Michaels, Seltzer and Bender also do not pay enough attention to how Dreiser, Wharton, London and Adams tested and developed evolutionary ideas. Studies of scientific observation and social environments have been restricted primarily to nineteenth-century American Romantic literature, for instance, Andrew McMurry, *Environmental Renaissance: Emerson, Thoreau, and the Systems of Nature* (Athens: University of Georgia Press, 2003).

6 Studies of naturalism as a genre of US literature relevant to this discussion are Charles C. Walcutt, *American Literary Naturalism: A Divided Stream* (Minneapolis: University of Minnesota Press, 1956); John J. Conder, *Naturalism in American Fiction: The Classic Phase* (Lexington: University of Kentucky Press, 1984); Lee Clark Mitchell, *Determined Fictions: American Literary Naturalism* (New York: Columbia University Press, 1989); Donald Pizer, *The Theory and Practice of American Literary Naturalism: Selected Essays and Reviews* (Carbondale: Southern Illinois University Press, 1982); June Howard, *Form and History in American Literary Naturalism* (Chapel Hill: University of North Carolina Press, 1985); Eric J. Sundquist (ed.), *American Realism: New Essays* (Baltimore, MD: Johns Hopkins University Press, 1982).

7 Peter J. Bowler remarks: 'There is no point in the individual's trying to work harder or improving its habits, since however hard it tries, rivals who are congenitally better equipped will beat it in the end'. See

The Non-Darwinian Revolution (Baltimore, MD: Johns Hopkins University Press, 1988), p. 37. This idea will be further examined in chapter 5.

8 Humberto Maturana and Francisco Varela, *The Tree of Knowledge: The Biological Roots of Human Understanding*, revised edn, trans. Robert Paolucci, foreword by J. Z. Young (Boston: Shambhala, 1992).

9 Niklas Luhmann, *Art as a Social System*, trans. Eva M. Knodt (Stanford, CA: Stanford University Press, 2000).

Chapter 1: Popular Science, Evolution and Global Information Management

1 Henry M. Stanley, 'The development of Africa', *Scientific American*, 75/16 (17 October 1896), 298.

2 Ira Livingston, 'Complex visuality: the radical middleground', in Bruce Clarke and Mark B. N. Hansen (eds), *Emergence and Embodiment: New Essays on Second-Order Systems Theory* (Durham, NC and London: Duke University Press, 2009), p. 252.

3 Karen Barad, *Meeting the Universe Halfway: Quantum Physics and the Entanglement of Matter and Meaning* (Durham, NC and London: Duke University Press, 2007), p. 133.

4 James McKeen Cattell, 'The progress of science: the popularization of science', *Popular Science Monthly*, 72/19 (April 1908), 384.

5 In Cary Wolfe, *What Is Posthumanism?* (Minneapolis: University of Minnesota Press, 2010), p. 235.

6 Refer to Robert T. Lagemann, 'New light on old rays: N-rays', *American Journal of Physics*, 45/3 (March 1977), 281–344.

7 'The N-rays: are they real or illusory?', *Scientific American* (4 June 1904), 434.

8 Wolfe, *Posthumanism*, p. 274.

9 'N-rays: real or illusory?', 434.

10 Wolfe, *Posthumanism*, p. 116.

11 John M. Coulter, 'Public interest in research', *Popular Science Monthly*, 72/19 (August 1905), 306.

12 Ibid., 309, 310.

13 Ibid., 312.

14 Gordon Webster, 'America's intellectual product', *Popular Science Monthly*, 72/13 (March 1908), 201.

15 Ibid., 202.

16 Ibid., 202.

17 Ibid., 204, 205.

18 Ibid., 209.

19 Barad, *Universe*, pp. 167–8.
20 Wolfe, *Posthumanism*, p. 257.
21 Ever the sceptic, Timothy Morton dismisses Barad's argument: 'Quantum-theory utopianism – Hey look! My mind can influence matter! – is just the upside of an all-too-true embeddedness of dominating mind in dominated world'. See *Ecology Without Nature: Rethinking Environmental Aesthetics* (Cambridge, MA: Harvard University Press, 2009), p. 27.
22 Ira Livingston, *Between Science and Literature: An Introduction to Autopoetics* (Urbana, IL and Chicago: University of Illinois Press, 2006), pp. 136–7.
23 Raymond Williams, *Marxism and Literature* (Oxford and New York: Oxford University Press, 1977), pp. 121–7.
24 John Carlos Rowe, 'Adams's education in the age of imperialism', in Rowe (ed.), *New Essays on The Education of Henry Adams* (Cambridge: Cambridge University Press, 1996), p. 89.
25 Daniel G. Brinton, 'The aims of anthropology', *Popular Science Monthly*, 47/1 (November 1895), 59.
26 Ibid., 65.
27 Ibid., 65.
28 Ibid., 69, 70.
29 Ibid., 71.
30 Ibid., 69.
31 Ibid., 70.
32 Mark Seltzer, *Bodies and Machines* (New York and London: Routledge, 1992), p. 113.
33 Ibid., pp. 113–14.
34 Lewis K. Harley, 'Race mixture and national character', *Popular Science Monthly*, 47/1 (May 1895), 86.
35 Walter Benn Michaels, *Our America: Nativism, Modernism, and Pluralism* (Durham, NC: Duke University Press, 1997), p. 12.
36 Harley, 'Race mixture', 88.
37 'Our trade with Africa', *Scientific American*, 74/25 (20 June 1896), 391.
38 Stanley, 'Development of Africa', 298.
39 Gustave Michaud, 'The climate of our new possessions', *Scientific American* (15 September 1900), 171.
40 Ibid.
41 M. W. Harrington, 'Natural products and resources of the Philippine Islands', *Scientific American*, 78/23 (4 June 1898), 355.
42 'The Philippine Islands', *Scientific American*, 78/19 (7 May 1898), 290.

43 Ibid.
44 'American trade with Central and South American countries', *Scientific American*, 78/19 (7 May 1898), 291.
45 Rowe, 'Imperialism', p. 92.
46 Ibid., p. 101.
47 Ibid., p. 109.
48 Niklas Luhmann, *Art as a Social System*, trans. Eva M. Knodt (Stanford, CA: Stanford University Press, 2000), p. 69.
49 Cary Wolfe, *Critical Environments: Postmodern Theory and the Pragmatics of the 'Outside'* (Minneapolis: University of Minnesota Press, 1998), p. 77.
50 Livingston, *Science and Literature*, p. 137.
51 Orson Desaix Munn, 'Political crisis', *Scientific American*, 73/34 (28 December 1895), 402.
52 Norman Angell, *The Great Illusion: A Study of the Relation of Military Power to National Advantage*, 4th edn (New York and London: G. P. Putnam's Sons, 1913).
53 Livingston, 'Complex visuality', p. 255.
54 Livingston, *Science and Literature*, p. 137.
55 Immanuel Wallerstein, *World Systems Analysis: An Introduction* (Durham, NC and London: Duke University Press, 2004), p. 2.
56 Ibid., p. 75, emphasis added. For a thoughtful application of world systems theory to literature, see Joseph Tabbi, 'World systems colliding: Thomas Pynchon and Niklas Luhmann', in Hannes Bergthaller and Carsten Schinko (eds), *Addressing Modernity: Social Systems Theory and US Cultures* (Amsterdam and New York: Rodopi, 2011), pp. 313–42.
57 Seltzer, *Bodies and Machines*, p. 159.
58 James Beniger, *The Control Revolution: Technological and Economic Origins of the Information Society* (Cambridge, MA: Harvard University Press, 1986), p. 32.
59 Ibid., p. 37.
60 In Beniger, *Control Revolution*, p. 38.
61 Ibid., p. 41.
62 Ibid., p. 42.
63 Ibid., p. 48. Maxwell's demon was a thought experiment. Maxwell imagined a miniscule entity who, by opening and closing a frictionless door, could sort molecules. The demon would let pass the faster molecules and block the slower ones. Consequently, the demon would reverse the process of entropy, obtaining energy without work.
64 Ibid., pp. 48, 49.

65 Ibid., p. 59.
66 Ibid., p. 66.
67 Wolfe, *Posthumanism*, p. 224.
68 Beniger, *Control Revolution*, pp. 78–89.
69 Erich Jantsch, *The Self-Organizing Universe: Scientific and Human Implications of the Emerging Paradigm of Evolution* (New York: Pergamon, 1980), p. 87.
70 Rowe, 'Imperialism', p. 107.
71 Henry Adams, *The Education of Henry Adams*, in *Adams: Novels, Mont Saint Michel, The Education*, ed. Ernest Samuels and Jayne N. Samuels (New York: Library of America, 1983), p. 1133.
72 Ibid.
73 Livingston, 'Complex visuality', p. 257.
74 Ibid.
75 Wolfe, *Posthumanism*, p. 222.
76 Barad, *Universe*, p. 171.
77 Henry Fairfield Osborn, 'Life and works of Darwin', *Popular Science Monthly*, 74 (April 1909), 335.
78 Ibid., 336.
79 Livingston, 'Complex visuality', p. 247.

Chapter 2: Dirty Naturalism and the Regime of Thermodynamic Self-Organization

1 Herbert Spencer, 'Professional institutions. IV. Orator and poet, actor and dramatist', *Popular Science Monthly*, 47/35 (August 1895), 434.
2 Spencer, 'Professional institutions. XI. Painter', *Popular Science Monthly*, 48/41 (March 1896), 704.
3 John Quincy Adams, 'Science versus art-appreciation', *Popular Science Monthly*, 62/25 (March 1903), 453.
4 Ibid., 454.
5 Ibid., 455.
6 Ibid., 462.
7 Ira Woods Howerth, 'The classification of the arts', *Popular Science Monthly*, 70/25 (May 1907), 431.
8 Ibid., 435.
9 Ibid., 436.
10 Jennifer L. Fleissner, *Women, Compulsion, Modernity: The Moment of American Naturalism* (Chicago: University of Chicago Press, 2004), pp. 161–200.

11 Theodore Dreiser, *Sister Carrie*, in *Sister Carrie, Jennie Gerhardt, Twelve Men*, ed. Richard Lehan (New York: Library of America, 1987), p. 442. For Dreiser's scientific background, see Louis J. Zanine, *Mechanism and Mysticism: The Influence of Science on the Thought and Work of Theodore Dreiser* (Philadelphia: University of Pennsylvania Press, 1993); Ronald E. Martin, *American Literature and the Universe of Force* (Durham, NC: Duke University Press, 1981).
12 Dreiser, *Sister Carrie*, p. 29.
13 Ibid., p. 46.
14 Ibid., p. 94.
15 Ibid., p. 94.
16 Ibid., p. 127.
17 Cary Wolfe, *What Is Posthumanism?* (Minneapolis: University of Minnesota Press, 2010), p. 234.
18 Niklas Luhmann, *Art as a Social System*, trans. Eva M. Knodt (Stanford, CA: Stanford University Press, 2000), p. 47.
19 Dreiser, *Sister Carrie*, p. 144.
20 Ibid., p. 403.
21 Ibid.
22 Judith Butler, *Bodies That Matter: On the Discursive Limits of 'Sex'* (New York and London: Routledge, 1993), pp. 12, 226.
23 Ibid., p. 226.
24 Dreiser, *Sister Carrie*, p. 144.
25 Wolfe, *Posthumanism*, p. 224.
26 Dreiser, *Sister Carrie*, pp. 141–2.
27 Ibid., p. 404.
28 Ibid., p. 447.
29 Ibid., p. 46.
30 Ibid., p. 404.
31 Ibid., p. 439.
32 Walter Benn Michaels, *The Gold Standard and the Logic of Naturalism: American Literature at the Turn of the Century* (Berkeley and London: University of California Press, 1987), pp. 41–5, 48–58.
33 Here I am playing on Michaels's idea that 'blood is blood, but – because blood is blood – blood isn't enough'. See *Our America: Nativism, Modernism, and Pluralism* (Durham, NC: Duke University Press, 1997), p. 12. For a discussion of blood as a significant motif in nineteenth-century US popular fiction, see Dale M. Bauer, '"In the blood": sentiment, sex, and the ugly girl', *differences*, 11/3 (1999), 57–75.

34 Niklas Luhmann, *Ecological Communication*, trans. John Bednarz, Jr (Chicago: University of Chicago Press, 1989), p. 16.

35 Wolfe, *Posthumanism*, p. 15.

36 Dreiser, *Sister Carrie*, p. 435.

37 Edith Wharton, *The House of Mirth*, in *Edith Wharton: Novels*, ed. R. W. B. Lewis (New York: Library of America, 1985), p. 297. For Wharton's reading of the evolutionary debates of the day, including works by Charles Darwin, T. H. Huxley, Herbert Spencer, Jean-Baptiste Lamarck, Alfred Russel Wallace and Ernst Haeckel, see her autobiography, *A Backwards Glance* (New York: D. Appleton-Century, 1934) and Shari Benstock, *No Gifts from Chance: A Biography of Edith Wharton* (New York: Scribner, 1994). Maureen Howard remarks: 'Wharton was playful in her use of the evolutionary metaphor in her fiction, yet also seriously committed to an understanding of evolution's complex relation to individual hereditary development and choice'. See '*The House of Mirth*: the bachelor and the baby', in Millicent Bell (ed.), *The Cambridge Companion to Edith Wharton* (Cambridge and New York: Cambridge University Press, 1995), p. 145.

38 Ibid., p. 301.

39 Ibid., p. 261, emphasis added.

40 Dale M. Bauer, 'Wharton's "others": addiction and intimacy', in Carol J. Singley (ed.), *A Historical Guide to Edith Wharton* (Oxford and New York: Oxford University Press, 2003), p. 120.

41 Elizabeth Ammons, 'Edith Wharton and race', in Bell, *Cambridge Companion to Edith Wharton*, pp. 70, 79–81. For further discussion of the novel's ending as a eugenic fantasy, refer to Jennie A. Kassanoff, *Edith Wharton and the Politics of Race* (Cambridge: Cambridge University Press, 2004). For a study of Wharton's later fiction and eugenics, see Dale M. Bauer, *Edith Wharton's Brave New Politics* (Madison: University of Wisconsin Press, 1994).

42 Wharton, *House of Mirth*, pp. 271, 204.

43 Nancy Bentley, 'Wharton, travel and modernity', in Singley, *Historical Guide to Edith Wharton*, pp. 147–79.

44 Cary Wolfe, *Animal Rites: American Culture, the Discourse of Species, and Posthumanist Theory* (Chicago: University of Chicago Press, 2003), p. 202.

45 Ibid.

46 Ammons, 'Race', p. 79.

47 Wharton, *House of Mirth*, p. 191.

48 Ibid., p. 244.

49 Ammons remarks that Wharton draws 'with ease on turn-of-the-century racializations that associated Jews with Asians'. See 'Race', p. 80. Thus Wharton's image of Rosedale's 'small sidelong eyes'(*House of Mirth*, p. 14) can be connected with Mrs Hatch's 'Oriental indolence'.
50 Wharton, *House of Mirth*, p. 290.
51 Ibid., p. 192.
52 Butler, *Bodies That Matter*, p. 12.
53 Wharton, *House of Mirth*, pp. 191, 196, 201.
54 Ibid., p. 208.
55 Ibid., p. 209.
56 Wolfe, *Posthumanism*, p. 237.
57 Ammons, 'Race', p. 78.
58 Wharton, *House of Mirth*, p. 135.
59 Ibid., p. 136.
60 Ibid.
61 Ibid., p. 138.
62 Ibid., p. 140.
63 Ibid.
64 Ibid., pp. 141–2. Refer to Ammons, 'Race', p. 79; Eleanor Dwight, 'Wharton and art', in Singley, *Historical Guide to Edith Wharton*, pp. 195–7; Emily J. Orlando, 'Picturing Lily: body art in *The House of Mirth*', in Gary Totten (ed.), *Memorial Boxes and Guarded Interiors: Edith Wharton and Material Culture* (Tuscaloosa: University of Alabama Press, 2007), pp. 83–110.
65 Wolfe, *Posthumanism*, p. 224.
66 Ammons comments: 'What Reynolds's *Mrs . Lloyd* stands for – nationalized Anglo-Saxon "natural" dominance – Lily embodies'. See 'Race', p. 79.
67 Wharton, *House of Mirth*, p. 142.
68 Ibid.
69 Ibid., p. 143.
70 Ibid., p. 146.
71 Wolfe, *Posthumanism*, p. 221.
72 Butler, *Bodies That Matter*, p. 12.
73 Bruce Clarke and Mark B. N. Hansen, 'Introduction: neocybernetic emergence', in Clarke and Hansen (eds), *Emergence and Embodiment: New Essays on Second-Order Systems Theory* (Durham, NC: Duke University Press, 2009), p. 9.
74 In Ira Livingston, *Between Science and Literature: An Introduction to Autopoetics* (Urbana, IL and Chicago: University of Illinois Press, 2006), p. 137.

75 Clarke and Hansen, 'Introduction', p. 9.

76 Livingston, *Science and Literature*, p. 140.

77 Erich Jantsch, *The Self-Organizing Universe: Scientific and Human Implications of the Emerging Paradigm of Evolution* (New York: Pergamon, 1980), p. 40.

78 Jeffrey S. Wicken, 'Thermodynamics, evolution, and emergence: ingredients for a new synthesis', in Bruce H. Weber, David J. Depew and James D. Smith (eds), *Entropy, Information, and Evolution: New Perspectives on Physical and Biological Evolution* (Cambridge, MA: MIT Press, 1988), pp. 146, 145.

79 Luhmann, *Ecological Communication*, p. 16.

80 In Livingston, *Science and Literature*, p. 137.

81 Livingston, *Science and Literature*, p. 138.

82 Ibid. Livingston adds: 'This breezy caricature overstates the hegemony of the new paradigms and understates the contradictions within and among them, but the breeziness is theirs, not mine'.

83 Ibid., p. 139.

84 Gavin Jones also addresses the theme of sinking in Lily's experience of poverty, represented by 'images of a self being submerged in abysmal depths of destitution', and her response to the moral 'weight' of her debt to Trenor. See *American Hungers: The Problem of Poverty in US Literature, 1840–1945* (Princeton, NJ and Oxford: Princeton University Press, 2008), pp. 95, 96–7. Fleissner suggests, however, that the narrative trajectories of Carrie and Lily do not follow the 'prevailing views of naturalism as a "plot of decline"', instead unfolding 'in a nonlinear fashion epitomized by a language of "drift"'. See *Women, Compulsion, Modernity*, p. 163.

85 Dreiser, *Sister Carrie*, p. 436.

86 Livingston, *Science and Literature*, p. 142.

87 Dreiser, *Sister Carrie*, p. 267.

88 Ibid., p. 442.

89 For Benetton's current advertisement campaigns, see 'Image gallery: brand campaigns' at their corporate website, *http://www.benettongroup.com/media-press/image-gallery/brand-campaigns*. For a collection of historical advertisements, see 'The fashion spot: The United Colors of Benetton – campaign archive' at *http://forums.thefashionspot.com/f79/united-colors-benetton-campaign-archive-77555.html*.

90 Also see Bentley, 'Wharton, travel and modernity', pp. 151–3.

91 Livingston, *Science and Literature*, p. 142.

92 Dreiser, *Sister Carrie*, p. 438.

93 Ibid., p. 403.

94 Livingston, *Science and Literature*, p. 137.

95 Ellen J. Goldner, 'The lying women and the cause of social anxiety: interdependence and the woman's body in *The House of Mirth*', *Women's Studies*, 21 (1992), 302.

96 Wharton, *House of Mirth*, pp. 141–2; Michaels, *Gold Standard*, pp. 240–1.

97 Fleissner, *Women, Compulsion, Modernity*, p. 190.

98 Luhmann, *Ecological Communication*, p. 19.

99 Ibid., p. 14.

100 Eric D. Schneider and Dorion Sagan, *Into the Cool: Energy Flow, Thermodynamics, and Life* (Chicago: University of Chicago Press, 2005), p. 6.

101 Ibid., p. 238.

102 Ibid., p. 240.

103 Ibid., p. 244.

104 C. Dyke, 'Cities as dissipative structures', in Weber, Depew and Smith, *Entropy*, p. 360.

105 Dreiser, *Sister Carrie*, p. 3.

106 Ibid., p. 15.

107 Dyke, 'Cities', p. 365.

108 Philipp Schweighauser, *The Noises of American Literature, 1890–1985: Toward a History of Literary Acoustics* (Gainesville, FL: University of Florida Press, 2006), p. 44.

109 Wharton, *House of Mirth*, p. 127.

110 Wai Chee Dimock, 'Debasing exchange: Edith Wharton's *The House of Mirth*', *PMLA*, 10/5 (1985), 787.

111 Dreiser, *Sister Carrie*, p. 88.

112 Ibid., p. 291.

113 Wharton, *House of Mirth*, p. 246.

114 Dyke, 'Cities', p. 365.

115 Michaels, *Gold Standard*, p. 228.

116 Schweighauser, *Noises of American Literature,* p. 47.

117 Ibid., p. 48.

118 William Rasch, 'Immanent systems, transcendental temptations, and the limits of ethics', in William Rasch and Cary Wolfe (eds), *Observing Complexity: Systems Theory and Postmodernity* (Minneapolis: University of Minnesota Press, 2000), p. 91.

119 Ibid., p. 93.

120 Wolfe, *Animal Rites*, p. 201.

121 Wharton, *House of Mirth*, p. 117.

122 Ibid., p. 140. Note that when Lily tries to talk about her financial 'worries' to her aunt (p. 179), her aunt responds negatively, 'shutting her lips with

the snap of a purse closing against a beggar'. This image is referenced by Fleissner, *Women, Compulsion, Modernity*, p. 194.

123 Ibid.

124 Ibid., p. 158.

125 Ibid., p. 159.

126 Ibid., p. 174.

127 Ibid., p. 175.

128 Dreiser, *Sister Carrie*, p. 3.

129 Wharton, *House of Mirth*, p. 159.

130 Ibid., p. 339. Gavin Jones challenges Dimock's assertion that Nettie is an idealized figure by arguing that she is instead a 'survivor', 'the final piece in Wharton's extensive examination of the psychology and culture of poverty, a piece that gives ultimate clarity to the conjunction of class and gender in *The House of Mirth*'. See *American Hungers*, p. 101.

131 Wharton, *House of Mirth*, pp. 340–1. On the ending, Fleissner posits: 'Lily gets hastily handed a baby, a kitchen, and a repudiation of drifting rootlessness . . . precisely because, in the book's first half, she might otherwise seem as ideally suited as any other urban "woman adrift" to *embody* the very malaises of modernity that she comes in her final revelation to diagnose'. See *Women, Compulsion, Modernity*, p. 196.

132 Wharton, *House of Mirth*, p. 347.

133 Goldner, 'Lying women', 286.

134 Wolfe, *Posthumanism*, pp. 232–5.

135 Dreiser, *Sister Carrie*, p. 423.

136 Jones, *American Hungers*, p. 65.

137 Ibid.

138 Ibid., p. 85. Bauer also addresses Lily's addiction; see 'Wharton's "others"', p. 121. Howard likewise connects the outcomes of Hurstwood and Lily; see 'Bachelor and the baby', p. 152.

139 Jones, *American Hungers*, p. 105.

140 Livingston, *Science and Literature*, p. 142.

141 Livingston, 'Complex visuality: the radical middleground', in Clarke and Hansen, *Emergence and Embodiment*, p. 261.

142 Jones, *American Hungers*, p. 85.

Chapter 3: The Ecology of Empire

1 Edward L. Thorndike, 'The evolution of the human intellect', *Popular Science Monthly*, 50/60 (November 1901), 59.

2 Ibid., 59.

3 Ibid., 60.

4 Ibid., 65.

5 Ibid.

6 Earle Labor, Robert C. Leitz III and I. Milo Shepard (eds), *The Letters of Jack London, vol. I: 1896–1905* (Stanford, CA: Stanford University Press, 1988), p. 489. I would like to thank Christopher Phelps for his valuable comments on the letters.

7 Ibid., p. 493.

8 Jack London, 'All Gold Canyon' , in *Jack London: Novels and Stories*, ed. Donald Pizer (New York: Library of America, 1982), pp. 777–96. Quoting from Earle Labor, Michael L. Johnson notes that the story exhibits London's 'newly awakened ecological conscience'. See Johnson, *Hunger for the Wild: America's Obsession with the Untamed West* (Lawrence: University of Kansas Press, 2007), p. 134.

9 Judith Butler, *The Psychic Life of Power: Theories in Subjection* (Stanford, CA: Stanford University Press, 1997), p. 30.

10 Richard Slotkin, *Gunfighter Nation: The Myth of the Frontier in Twentieth-Century America* (New York: Athenaeum, 1992), p. 4.

11 Ibid., pp. 30–1.

12 Theodore Roosevelt, 'The Strenuous Life', in *The Works of Theodore Roosevelt, vol. XV* (New York: Scribner, 1923–6), pp. 270–1. For Mark Seltzer, the speech contributes to the transformation of 'competitive individualism' to 'disciplinary individualism'; see *Bodies and Machines* (New York and London: Routledge, 1992), p. 155. While London's dog novels reflect this transformation, such a change seems in the end only partial, given London's overall ambivalence about what Seltzer calls 'machine culture'.

13 Jennifer L. Fleissner, *Women, Compulsion, Modernity: The Moment of American Naturalism* (Chicago: University of Chicago Press, 2004), p. 83.

14 London, *The Call of the Wild*, in *Novels and Stories*, pp. 5–86, p. 5. All subsequent citations from *The Call of the Wild* and *White Fang* refer to this edition.

15 Ibid., p. 6.

16 Slotkin, *Gunfighter Nation*, pp. 34–5.

17 London, *Call of the Wild*, p. 6.

18 Vicki Hearne, *Adam's Task: Calling Animals by Name* (New York: Random House, 1987), p. 48.

19 John Myers O'Hara's poem was published in *The Bookman*, 16 (November 1902). See London, *Novels and Stories*, p. 1007.

20 Thomas Henry Huxley, *Evolution and Ethics: The Romanes Lecture* (New

York: D. Appleton & Company, 1894), pp. 80–1. London's reading of evolutionary theorists can be found in David Mike Hamilton, '*The Tools of My Trade*': *The Annotated Books in Jack London's Library* (Seattle: University of Washington Press, 1986) and Labor et al., *The Letters of Jack London* (3 vols). All subsequent citations from the *Letters* refer to this edition.

21 The idea that London's dog narratives construct evolutionary experiments is taken from Doug Davis.

22 Lawrence I. Berkove, 'Jack London and evolution: from Spencer to Huxley', *American Literary Realism*, 36/3 (Spring 2004), 246. See in particular his discussion of London's story, 'The Law of Life', 249.

23 *Letters of Jack London, vol. I*, pp. 282–3.

24 Seltzer, *Bodies and Machines*, p. 163.

25 London, *Call of the Wild*, p. 12.

26 Cary Wolfe reads Ernest Hemingway's *The Sun Also Rises* as an example of how male bonding is produced through the physical sacrifice of animal bodies in bullfighting. Refer to *Animal Rites: American Culture, the Discourse of Species, and Posthumanist Theory* (Chicago: University of Chicago Press, 2003), p. 138. I am interested in how London's codes of masculine discipline, inscribed and sustained through the bodies of dogs, come before the 'species discourse' of Hemingway. Wolfe uses this idea to explore the determining of human–animal relations in the novel.

27 London, *Call of the Wild*, p. 15.

28 Ibid., p. 16.

29 *Letters of Jack London, vol. I*, pp. 88–9.

30 Refer to Carol J. Adams, 'The feminist traffic in animals', in Greta Gaard (ed.), *Ecofeminism: Women, Animals, Nature* (Philadelphia: Temple University Press, 1993), pp. 197–206.

31 London, *Call of the Wild*, p. 22.

32 Ibid.

33 Ibid., p. 23.

34 Ibid., p. 25.

35 *Letters of Jack London, vol. I*, p. 89.

36 London, *Call of the Wild*, p. 29.

37 Judith Butler, *Bodies That Matter: On the Discursive Limits of 'Sex'* (New York: Routledge, 1993), p. 12.

38 Jonathan Auerbach, *Male Call: Becoming Jack London* (Durham, NC: Duke University Press, 1996), pp. 84–114. While Auerbach offers a strong reading of *The Call of the Wild* through London's development as a professional writer, he persistently denies the validity of London's

evolutionary thinking, which, I argue, has historical, scientific and theoretical significance.

39 London, *Call of the Wild*, p. 30.
40 Auerbach, *Male Call*, p. 106.
41 London, *Call of the Wild*, p. 52.
42 Ibid., p. 46.
43 Ibid., pp. 46, 52.
44 Auerbach, *Male Call*, p. 104.
45 London, *Call of the Wild*, p. 55.
46 Ibid., pp. 54, 60.
47 Ibid., p. 59.
48 Ibid., pp. 60–1.
49 Ibid., p. 68.
50 John F. Kasson, *Houdini, Tarzan, and the Perfect Man: The White Male Body and The Challenge of Modernity in America* (New York: Hill and Wang, 2001), p. 38. Kasson repeatedly emphasizes how Sandow performs for audiences, 'showing off' his body.
51 London, *Call of the Wild*, p. 69; Kasson, *Perfect Man*, pp. 54, 35.
52 London, *Call of the Wild*, p. 67; Kasson, *Perfect Man*, p. 29.
53 London, *Call of the Wild*, p. 70.
54 London, *White Fang*, in *Novels and Stories*, pp. 91–284, p. 92.
55 Seltzer, *Bodies and Machines*, p. 166.
56 Slotkin, *Gunfighter Nation*, p. 35.
57 London, *White Fang*, p. 121.
58 Ibid., pp. 122–3.
59 Ibid., p. 173.
60 Ibid., p. 176.
61 Ibid., p. 185.
62 Ibid., p. 189.
63 Kasson, *Perfect Man*, p. 54.
64 London, *White Fang*, p. 209.
65 Ibid., p. 210.
66 Ibid., p. 234.
67 *Letters of Jack London, vol. I*, pp. 88–9.
68 London, *White Fang*, p. 237.
69 Ibid., p. 242.
70 Ibid., p. 243.
71 Ibid., p. 244.
72 Charles Darwin, *The Expression of Emotions in Man and Animals*, ed. Francis Darwin (London: William Pickering, 1989), p. 215.

73 Seltzer, *Bodies and Machines*, p. 169.
74 Marjorie Garber, *Dog Love* (New York: Simon & Schuster, 1996), p. 121.
75 Eve Kosofsky Sedgwick, *Tendencies* (Durham, NC: Duke University Press, 1993), p. 171.
76 London, *White Fang*, p. 250.
77 Ibid., p. 252.
78 Ibid., p. 255.
79 Slotkin, *Gunfighter Nation*, p. 63.
80 Kasson, *Perfect Man*, pp. 50, 52, 51.
81 London, *White Fang*, p. 255.
82 Ibid., p. 255.
83 Ibid., p. 264.
84 Lynn Margulis and Dorion Sagan, *What Is Life?* (New York: Simon & Schuster, 1995), p. 22. For a discussion of how their model of globally connected, autopoietic ecosystems (influenced by James Lovelock's Gaia hypothesis) argues against the idea of a liberal humanist subject, see my article, 'Becoming American: evolution and performance in Edith Wharton's *The Custom of the Country*', *Intertexts*, 9/1 (Spring 2005), 43–59.
85 Margulis and Sagan, *What is Life?*, p. 199.
86 Charles Darwin, *On the Origin of Species: A Facsimile of the First Edition* (Cambridge, MA: Harvard University Press, 1966), p. 489.
87 Darwin, *The Descent of Man and Selection in Relation to Sex*, 2nd ed. (Chicago: Rand, McNally & Co., 1974), p. 122.
88 London, 'The Other Animals', in Dale L. Walker and Jeanne Campbell Reesman (eds), *No Mentor But Myself: Jack London on Writing and Writers*, 2nd ed. (Stanford, CA: Stanford University Press, 1999), pp. 109–10.
89 For accounts of dog behaviour, London drew upon Egerton R. Young, *My Dogs in the Northland*, 5th ed. (New York: Fleming H. Revell & Co., 1902) and Edward Jesse, *Anecdotes of Dogs* (London: Bell & Daldy, 1873).
90 London, *Call of the Wild*, p. 21.
91 London, *White Fang*, p. 203.
92 Ibid., p. 269.
93 London, *Call of the Wild*, p. 41. London follows Darwin, who asserts that 'dogs, cats, horses, and probably all the higher animals, even birds, have vivid dreams'. See *Descent of Man*, p. 72. I will try to resolve the problem of how dreams can recall an evolutionary past later on in the chapter.
94 Donna J. Haraway, *Modest_Witness@Second Millennium. FemaleMan_Meets_OncoMouse* (New York and London: Routledge, 1997), pp. 199–200.
95 Donna J. Haraway, *The Companion Species Manifesto: Dogs, People, and Significant Otherness* (Chicago: Prickly Paradigm Press, 2003), p. 29.

For a discussion of Haraway's study of dog genetics and intelligence, refer to my article, 'Furry logic: biological kinship and empire in Jack London's *The Call of the Wild*', *Interdisciplinary Studies in Literature and Environment* (*ISLE*), 14/1 (Winter 2007), 25–49.

96 London, *White Fang*, p. 172.

97 Ibid., pp. 179, 182.

98 Ibid., p. 214.

99 London, 'The other animals', p. 120.

100 Humberto R. Maturana and Francisco J. Varela, *The Tree of Knowledge: The Biological Roots of Human Understanding* (Boston and London: Shambhala, 1987), p. 45.

101 London, *Call of the Wild*, p. 62.

102 London, *White Fang*, p. 155.

103 Donna J. Haraway, *Simians, Cyborgs, and Women: The Reinvention of Nature* (London and New York: Routledge, 1991), pp. 199–200.

104 Wolfe, *Animal Rites*, pp. 203–4.

105 Ibid., p. 207.

106 Jacques Derrida, 'The animal that therefore I am (more to follow)', trans. David Wills, *Critical Inquiry*, 28 (Winter 2008), 415.

107 Wolfe, *Animal Rites*, p. 66.

108 London, *Call of the Wild*, p. 22.

109 In Lisa Hopkins, 'Jack London's evolutionary hierarchies: dogs, wolves, and men', in Lois A. Cuddy and Claire M. Roche (eds), *Evolution and Eugenics in American Literature and Culture, 1880–1940: Essays on Ideological Conflict and Complicity* (Lewisburg, PA: Bucknell University Press, 2003), p. 62.

110 London, *White Fang*, pp. 194–5.

111 Ibid., p. 263.

112 London, 'The Other Animals', p. 120.

113 Ibid., p. 119.

114 Annette Kolodny, *The Lay of the Land: Metaphor as Experience and History in American Life and Letters* (Chapel Hill: University of North Carolina Press, 1975), p. 133.

115 London, *Call of the Wild*, p. 66.

116 Evelyn Fox Keller, 'From secrets of life to secrets of death', in Keller et al. (eds), *Three Cultures: Fifteen Lectures on the Confrontation of Academic Cultures* (The Hague: Universitaire Pers Rotterdam, 1989), p. 4. Also see Stacy Alaimo, *Undomesticated Ground: Recasting Nature as Feminist Space* (Ithaca, NY and London: Cornell University Press, 2000).

117 Johnson, *Hunger for the Wild*, p. 38.

118 Ibid., p. 131.

119 London, *Call of the Wild*, pp. 67, 69.

120 Auerbach, *Male Call*, p. 107.

121 London, *Call of the Wild*, p. 71.

122 Ibid., p. 72.

123 London critiques the current version of the 'traffic in animals': 'the production, transportation, slaughter, and packaging of animals' bodies' for human consumption. See 'Animals', p. 197.

124 London, *Call of the Wild*, p. 61.

125 Auerbach, *Male Call*, p. 106.

126 London, *Call of the Wild*, p. 76.

127 Kasson, *Perfect Man*, pp. 23–8. Butler questions 'whether parodying the dominant norms is enough to displace them; indeed, whether the denaturalization of gender cannot be the very vehicle for a reconsolidation of hegemonic norms'. Refer to *Bodies That Matter*, p. 125.

128 London, *Call of the Wild*, pp. 77–8.

129 Kasson, *Perfect Man*, pp. 68–9.

130 Reading Roosevelt's account of his own life, Kasson discusses how he remade 'his "sickly, delicate," and asthmatic body into the strong, vigorous frame of a fearless leader'. See *Perfect Man*, pp. 30–1.

131 Slotkin, *Gunfighter Nation*, pp. 52–3. For a reading of the end of *The Call of the Wild* as a justification for US imperialism, see Christopher Gair, *Complicity and Resistance in Jack London's Novels* (Lewiston, NY: Edwin Mellon, 1997), pp. 56–7, 67; Amy Kaplan, 'Nation, region, and empire', in Emory Elliot (ed.), *Columbia History of the American Novel* (New York: Columbia University Press, 1991), p. 265. For an interpretation of Roosevelt's speech as social Darwinist doctrine, see Gail Bederman, *Manliness and Civilization: A Cultural History of Gender and Race in the United States, 1880–1917* (Chicago: University of Chicago Press, 1995). John Pettegrew considers *The Call of the Wild* and Owen Wister's *The Virginian* as masculine frontier narratives. See *Brutes in Suits: Male Sensibility in America, 1890–1920* (Baltimore, MD: Johns Hopkins University Press, 2007), pp. 94–111.

132 Refer to Shelley Streeby, *American Sensations: Class, Empire, and the Production of Popular Culture* (Berkeley, CA, Los Angeles and London: University of California Press, 2002).

133 For a discussion of the 'vanishing Indian', see Johnson, *Hunger for the Wild*, pp. 192–204; Alan Trachtenberg, *Shades of Hiawatha: Staging Indians, Making Americans, 1880–1930* (New York: Hill and Wang, 2004).

134 London, *Call of the Wild*, p. 85. Joan Hedrick also examines Buck's apparent escape from captivity. See *Solitary Comrade: Jack London and His Work* (Chapel Hill: University of North Carolina Press, 1982), pp. 94–111.

135 Kasson, *Perfect Man*, p. 54.

136 Ibid., pp. 80–1.

137 Ibid., p. 92.

138 Ibid., p. 116.

139 Ibid., p. 155.

140 Priscilla Wald, *Constituting Americans: Cultural Anxiety and Narrative Form* (Durham, NC and London: Duke University Press, 1995), p. 5.

141 Ibid., p. 7.

142 Ibid., p. 16.

143 Ibid., p. 224.

144 Cited in Wald, *Constituting Americans*, p. 224.

145 London, *White Fang*, p. 204.

146 Ibid., p. 231.

147 Slotkin, *Gunfighter Nation*, p. 39.

148 Haraway, *Companion Species Manifesto*, p. 13.

149 Greta Gaard, 'Living interconnections with animals and nature', in Gaard, *Ecofeminism*, p. 5.

150 London, *White Fang*, p. 277.

151 Ibid., p. 278.

152 Ibid., p. 279.

153 Ibid., p. 280. Kaplan notes that both novels end in scenes of violence. See 'Nation, region, and empire', p. 265.

154 Ibid., p. 283.

155 Ibid., p. 282.

156 Ibid., p. 268.

157 Butler, *Bodies That Matter*, p. 12.

158 Auerbach, *Male Call*, p. 229.

159 Butler, *Bodies That Matter*, p. 12.

160 Butler, *The Psychic Life of Power*, p. 30.

161 *Letters of Jack London, vol. 1*, p. 89.

162 Ibid., p. 92.

163 John Howard Moore, *Universal Kinship* (Chicago: Charles H. Kerr & Co., 1906), p. 324. London's personal copy is housed in the Huntington Library Rare Book Collection, San Marino, CA. I would like to thank Sara S. Hodson and the rest of the library staff for their help with my research on London.

164 Wolfe, *Animal Rites*, p. 90.

Chapter 4: After the Flood: Performance and Nation

1 George C. Whipple, 'Municipal water-works laboratories', *Popular Science Monthly* (December 1900), 172.
2 Ibid., 173.
3 Ibid., 179.
4 Ibid., 180.
5 Ibid., 180–1.
6 Ibid., 182.
7 See Stacy Alaimo, *Bodily Natures: Science, Environment, and the Material Self* (Bloomington and Indianapolis: Indiana University Press, 2010), pp. 64–76.
8 Cecelia Tichi, 'Emerson, Darwin, and *The Custom of the Country*', in Carol J. Singley (ed.), *A Historical Guide to Edith Wharton* (Oxford: Oxford University Press, 2003), p. 95.
9 Edith Wharton, *The Custom of the Country*, in *Edith Wharton: Novels*, ed. R. W. B. Lewis (New York: Library of America, 1985), p. 675. All subsequent citations from the story refer to this volume.
10 Ibid.
11 Ibid.
12 Ibid., p. 778.
13 Carol B. Sapora, 'Undine Spragg, the mirror and the lamp in *The Custom of the Country*', in Gary Totten (ed.), *Memorial Boxes and Guarded Interiors: Edith Wharton and Material Culture* (Tuscaloosa: University of Alabama Press, 2007), p. 280.
14 Wharton, *Custom of the Country*, p. 771.
15 Clare Preston, *Edith Wharton's Social Register* (New York: St. Martin's Press, 2000), p. 108.
16 Wharton, *Custom of the Country*, p. 922.
17 Ibid., p. 925.
18 Michel Foucault, 'Right of death and power over life', in Paul Rabinow (ed.), *The Foucault Reader* (New York: Pantheon, 1984), p. 264.
19 Ibid., p. 268.
20 Ibid., p. 266.
21 Cary Wolfe, *What Is Posthumanism?* (Minneapolis: University of Minnesota Press, 2010), p. 53.
22 Mark Seltzer, *Bodies and Machines* (New York and London: Routledge, 1992), p. 164.
23 Wharton, *Custom of the Country*, p. 928.
24 Preston, *Social Register*, p. 108.

25 In Wolfe, *Posthumanism*, p. 56.

26 Wharton, *Custom of the Country*, p. 779.

27 Ibid., p. 751.

28 Ibid., p. 745.

29 Ibid., p. 750.

30 Nancy Bentley, *The Ethnography of Manners: Hawthorne, James, Wharton* (Oxford: Oxford University Press, 2003), p. 197.

31 Wharton, *Custom of the Country*, p. 658.

32 Richard W. Etulain, *Beyond the Missouri: The Story of the American West* (Albuquerque: University of New Mexico Press, 2006), pp. 300–1. Thanks to Janet Sharistanian for the suggestion of Kansas City as a model for the fictional Apex City. The evidence can be found in a telling image of Undine's anger: 'Everything had gone down before her, as towns and villages went down before one of the tornadoes of her native state'. See Wharton, *Custom of the Country*, p. 703. It seems fitting that Mike Davis labels his chapter on the overlooked frequency of tornadoes in Los Angeles, 'Our secret Kansas'. Refer to *Ecology of Fear: Los Angeles and the Imagination of Disaster* (New York: Metropolitan Books, 1998).

33 Wharton, *Custom of the Country*, p. 982.

34 Ibid., p. 669.

35 Ibid., p. 846. For an interpretation of Ralph's fantasy as a non-satirical reflection of Wharton's social class, see Jennie A. Kassanoff, *Edith Wharton and the Politics of Race* (Cambridge: Cambridge University Press, 2004). Wharton's reading of early-twentieth-century studies of evolution by Vernon L. Kellogg and Robert H. Lock has been treated by Tichi, 'Emerson, Darwin, and *The Custom of the Country*' and Martha H. Patterson, 'Incorporating the new woman in Wharton's *The Custom of the Country*', *Studies in American Fiction*, 26/2 (1998), 213–36. For Tichi, Undine's son, Paul, represents 'the regeneration and revitalization' (111) of Wharton's class. Patterson sees Wharton's novel as restaging the debate between Kellogg, who critiqued natural selection, and Lock, who favoured a neo-Darwinian doctrine of eugenics. She then suggests that 'incorporation . . . supersedes eugenics as a means of social control' (221).

36 Michael L. Johnson, *Hunger for the Wild: America's Obsession with the Untamed West* (Lawrence: University of Kansas Press, 2007), p. 188.

37 Ibid., p. 189.

38 Bentley, *Ethnography of Manners*, p. 192. Refer to Charlotte Perkins Gilman, *Woman and Economics* (New York: Harper & Row, 1966), p. 58.

39 Wharton, *Custom of the Country*, p. 1008.

40 Ibid., p. 793.

41 Ibid., p. 795.

42 Johnson uses here the idea developed by Renato Rosaldo, *Culture and Truth: The Remaking of Social Analysis* (Boston: Beacon Press, 1989), pp. 69, 70. See Johnson, *Hunger for the Wild*, p. 207.

43 Karen M. Keefer, 'Dirty water and clean toilets: medical aspects of the 1904 Louisiana Purchase Exposition', *Gateway Heritage*, 9/1 (Summer 1988), 33. Thanks to Pete Coco for suggesting all cited articles for the history of Western US water management.

44 Ibid., 34.

45 Carolyn G. Shapiro-Shapin, '"A really excellent scientific contribution": scientific creativity, scientific professionalism, and the Chicago Drainage Case, 1900–1906', *Bulletin of the History of Medicine*, 71/3 (1997), 390.

46 Ibid., 391.

47 T. D. Samuel, Jr, 'The water supply system of Kansas City, Missouri', *American Water Works Association Journal,* 22/9 (1930), 1236.

48 James Earl Sherow, 'The art of water and the art of living', *Kansas History*, 25/1 (2002), 58.

49 Ibid., 60.

50 Amahia Mallea, 'Downstreamers: public health and relationships on the Missouri River', *Agricultural History*, 76/2 (Spring 2002), 393.

51 Ibid., 394.

52 Ibid., 396.

53 Ibid.

54 Ibid., 397.

55 Ibid., 398.

56 Ibid.

57 Ibid., 400.

58 Ibid., 401.

59 Ibid.

60 Ibid., 403.

61 Ibid.

62 Wharton, *Custom of the Country*, p. 749.

63 Cary Wolfe, *Animal Rites: American Culture, the Discourse of Species, and Posthumanist Theory* (Chicago and London: University of Chicago Press, 2003), p. 202.

64 Ibid., p. 202.

65 Wharton, *Custom of the Country*, p. 625.

66 Judith Butler, *Bodies That Matter: On the Discursive Limits of 'Sex'* (New York and London: Routledge, 1993), pp. 12, 226.

67 Wharton, *Custom of the Country*, p. 727.

68 Ibid., p. 633. The issue of authenticity in the novel is evaluated by Christopher Gair, 'The crumbling structure of "appearances": representation and authenticity in *The House of Mirth* and *The Custom of the Country*', *Modern Fiction Studies*, 43/2 (1997), 349–73; Ticien Marie Sassoubre, 'Property and identity in *The Custom of the Country*', *Modern Fiction Studies*, 49/4 (2003), 687–713.
69 Wharton, *Custom of the Country*, p. 688.
70 Butler, *Bodies that Matter*, p. 226.
71 Wharton, *Custom of the Country*, p. 673.
72 Ibid., p. 939.
73 Butler, *Bodies That Matter*, p. 12.
74 Bentley, *Ethnography of Manners*, p. 203.
75 Wharton, *Custom of the Country*, p. 806.
76 Butler, *Bodies That Matter*, p. 224.
77 Bentley, *Ethnography of Manners*, p. 164.
78 Ibid., p. 185.
79 Wharton, *Custom of the Country*, pp. 802, 807.
80 Wharton, *Custom of the Country*, p. 802.
81 Bentley, *Ethnography of Manners*, p. 183.
82 Thanks to Lisa Yaszek for pointing this idea out to me.
83 Wharton, *Custom of the Country*, p. 795.
84 Ibid., p. 635.
85 Ibid., p. 879.
86 Bentley, *Ethnography of Manners*, pp. 196, 187.
87 Priscilla Wald, *Constituting Americans: Cultural Anxiety and Narrative Form* (Durham, NC and London: Duke University Press, 1995), p. 5.
88 Ibid., p. 7.
89 Wharton, *Custom of the Country*, p. 853.
90 Ibid., pp. 802–3.
91 Roland Robertson, *Globalization: Social Theory and Global Culture* (London: Sage, 1992), pp. 29, 59. Also referenced by Bentley, 'Wharton, travel, and modernity', in Singley, *Historical Guide to Edith Wharton*, p. 149.
92 Archibald Cary Coolidge, *The United States as a World Power* (New York: Macmillan, 1909), p. 40.
93 Ibid., p. 44.
94 William Morton Fullerton, *Problems of Power: A Study of International Politics from Sadowa to Kirk-Kilisse* (London: Constable & Company Ltd, 1913), p. 10.
95 Ibid., p. 8.

96 For opposing perspectives of Coolidge and Fullerton's influence on Wharton's thinking, see Dale M. Bauer, *Edith Wharton's Brave New Politics* (Madison: University of Wisconsin Press, 1994); Frederick Wegener, '"Rabid imperialist": Edith Wharton and the obligations of empire in modern American fiction', *American Literature*, 72/4 (December 2000), 783–812.
97 Wharton, *Custom of the Country*, pp. 728, 966.
98 Barry Shank, 'Iterating the collective subject: performative history and the ethics of cultural politics' (unpublished paper presented at the Performance Seminar, Hall Center for the Humanities, University of Kansas, Autumn 1998). I would like to thank the author for permission to quote from his paper.
99 Wegener, '"Rabid imperialist"', 786.
100 Coolidge, *World Power*, p. 62.
101 Ibid., p. 63.
102 Wald, *Constituting Americans*, p. 16.
103 Ibid., p. 224.
104 Butler, *Bodies That Matter*, pp. 225–6.
105 Ibid., pp. 226–7.
106 Refer to Philip J. Deloria, *Playing Indian* (New Haven, CT and London: Yale University Press, 1998).
107 In Wald, *Constituting Americans*, p. 244.
108 Wald, *Constituting Americans*, p. 244.
109 Wharton, *Custom of the Country*, p. 951.
110 In Wald, *Constituting Americans*, p. 245.
111 Lauren Berlant, *The Anatomy of National Fantasy: Hawthorne, Utopia, and Everyday Life* (Chicago: University of Chicago Press, 1991), p. 5.
112 Karen Barad, *Meeting the Universe Halfway: Quantum Physics and the Entanglement of Matter and Meaning* (Durham, NC and New York: Duke University Press, 2007), pp. 145–6.
113 Wolfe, *Posthumanism*, pp. 53, 56.
114 Barad, *Universe*, p. 151.
115 Wharton, *Custom of the Country*, p. 745.
116 Bentley, *Ethnography of Manners*, p. 203.
117 Wolfe, *Posthumanism*, p. 222.
118 Niklas Luhmann, *Art as a Social System*, trans. Eva M. Knodt (Stanford, CA: Stanford University Press, 2000), p. 92.
119 Wharton, *Custom of the Country*, p. 720.
120 Ibid., p. 721.
121 Ibid., p. 724.

122 Ibid., p. 795.

123 Ibid., p. 903.

124 For a discussion of 'the art of writing' in the novel, see Bentley, *Ethnography of Manners*, p. 178. Tichi proposes that 'in Ralph, Wharton gives us a version of Emerson, but deliberately an Emerson manqué'. See 'Emerson, Darwin, and *The Custom of the Country*', p. 102.

125 Wharton, *Custom of the Country*, pp. 745, 748.

126 Ibid., p. 802, emphasis added.

127 Wolfe, *Posthumanism*, p. 224.

128 Luhmann, *Art as a Social System*, p. 69.

129 Wharton, *Custom of the Country*, p. 746.

130 Wolfe, *Posthumanism*, pp. 232–3.

131 Ibid., p. 234.

132 Luhmann, *Art as a Social System*, p. 47.

133 Wharton, *Custom of the Country*, p. 750.

134 Ibid., p. 750.

135 Bentley, *Ethnography of Manners*, p. 194; Wharton, *Custom of the Country*, p. 710. The portrait, as described by Wharton, is an allusion to J. W. Waterhouse's painting *Undine*.

136 Wharton, *Custom of the Country*, p. 752.

137 Ibid., p. 754.

138 Ibid., pp. 755, 754.

139 Ibid., p. 758.

140 Bentley, *Ethnography of Manners*, p. 205.

141 Wolfe, *Posthumanism*, p. 237.

142 Wolfe, *Posthumanism*, p. 94.

143 Bentley, *Ethnography of Manners*, p. 181; Wharton, *Custom of the Country*, p. 1007.

144 Wharton, *Custom of the Country*, p. 968.

145 Bentley explains: 'The fantasy of total social mobility is never without the accompaniment of a darker phantasm of Undine's "monstrous" potential for inflicting social damage'. See *Ethnography of Manners*, p. 206. Refer as well to Eugene Thacker, 'Necrologies, or death of the "body politic"', in Patricia T. Clough and Craig Willse (eds), *Beyond Biopolitics* (Durham, NC and New York: Duke University Press, 2011), p. 152. Thacker (pp. 146–51), extends the idea of the monstrous to the larger concept of the body politic. Also see Wolfe, *Before the Law: Humans and Other Animals in a Biopolitical Frame* (Chicago and London: University of Chicago Press, 2013), pp. 37–9.

146 Bentley, *Ethnography of Manners*, p. 210; Wharton, *Custom of the Country*, p. 1014.

147 Luhmann, *Ecological Communication*, trans. John Bednarz, Jr (Chicago: University of Chicago Press, 1989), p. 14.

148 Witness the attempt of my current state of residence to rebrand itself as 'Pure Michigan', effacing the industrial history, and legacy of pollution, of cities such as Detroit.

Chapter 5: The Miseducation of Henry Adams: Fantasies of Race, Citizenship and Biological Dynamos

1 W. S. Franklin, 'The Second Law of Thermodynamics: its basis in intuition and common sense', *Popular Science Monthly*, 76/15 (March 1910), 227.

2 Ibid., 227.

3 Ibid., 231.

4 Ibid., 232.

5 Ibid., 233.

6 Ibid., 236.

7 Thomas Pynchon discusses his reading of Henry Adams's *The Education of Henry Adams* in *Slow Learner* (Boston and Toronto: Little, Brown & Company, 1984), p. 13.

8 Thomas Pynchon, *Against the Day* (New York: Penguin, 2006), p. 10.

9 Charles Darwin, *On the Origin of Species: A Facsimile of the First Edition* (Cambridge, MA: Harvard University Press, 1966); *The Descent of Man and Selection in Relation to Sex* (New York: AMS Press, 1972).

10 Peter J. Bowler, *The Non-Darwinian Revolution* (Baltimore, MD: Johns Hopkins University Press, 1988), p. 145.

11 Darwin, *Descent*, p. 135.

12 Ibid., p. 141.

13 Bowler, *Non-Darwinian Revolution*, p. 37.

14 Darwin, *Descent*, p. 142.

15 Ibid., p. 565.

16 Ibid.

17 Darwin, *Descent*, p. 565.

18 Stephen Jay Gould, *Ever Since Darwin* (New York: W. W. Norton, 1977), p. 38.

19 Bowler, *Non-Darwinian Revolution*, p. 157.

20 Ibid., pp. 157, 179, 181. Bowler argues that Darwin's evolutionary theories do not gain acceptance until they are 'synthesized' with genetics later in the twentieth century. Gould, however, in *The Panda's Thumb: More Reflections in Natural History* (New York: W. W. Norton, 1980),

contends that this synthesis still leaves unresolved questions about the rate of change in evolution. Refer to the chapter, 'The episodic nature of evolutionary change', pp. 179–85.

21 Henry Adams, in *Adams: Novels, Mont Saint Michel, The Education*, ed. Ernest Samuels and Jayne N. Samuels (New York: Library of America, 1983), p. 925.

22 J. C. Levenson, 'Henry Adams, U. S. Grant, and evolution: practicing history in the age of Darwin', in William Merrill Decker and Earl N. Harbert (eds), *Henry Adams and the Need to Know* (Boston: Massachusetts Historical Society, 2005), pp. 357, 359. While Levenson gives the most detailed account of Adams's participation in evolutionary debates, other accounts may be found in William Merrill Decker, 'A martyr to the disease of omniscience', in Decker and Harbert, *The Need to Know*, pp. 318–44; N. Katherine Hayles, *Chaos Bound* (Ithaca, NY and London: Cornell University Press, 1990); Ronald E. Martin, *American Literature and the Universe of Force* (Durham, NC: Duke University Press, 1981).

23 Adams, *Education*, p. 926.

24 Levenson, 'Practicing history', pp. 356–7, 361, 363–4. Adams read Spencer's *Philosophy of Style*, which he found rather unconvincing.

25 Ibid., p. 358.

26 Adams, *Education*, p. 937.

27 Ibid., p. 938.

28 Alan Trachtenberg, *Shades of Hiawatha: Staging Indians, Making Americans, 1880–1930* (New York: Hill and Wang, 2004), p. 4.

29 Richard Slotkin, *Gunfighter Nation: The Myth of the Frontier in Twentieth-Century America* (New York: Athenaeum, 1992), p. 53.

30 For an investigation of how Native American identities were performed by non-Native Americans, refer to Philip J. Deloria, *Playing Indian* (New Haven, CT: Yale University Press, 1998).

31 Cindy Weinstein, 'From true woman to new woman to virgin', in Decker and Harbert, *The Need to Know*, p. 309.

32 Trachtenberg, *Shades of Hiawatha*, pp. 7–8.

33 Michael L. Johnson, *Hunger for the Wild: America's Obsession with the Untamed West* (Lawrence: University of Kansas Press), p. 207. Johnson uses the idea of 'imperialist nostalgia' found in Renato Rosaldo, *Culture and Truth: The Remaking of Social Analysis* (Boston: Beacon Press, 1989), pp. 69, 70.

34 Crosbie Smith and Ian Higginson, '"No traces of a beginning, no prospect of an end": Henry Adams, Charles Lyell, and the politics of uniformity', in Decker and Harbert, *The Need to Know*, p. 114.

35 Adams, *Education*, p. 1028.
36 Decker, 'A martyr', p. 331.
37 Slotkin, *Gunfighter Nation*, p. 63.
38 Adams, *Education*, p. 1031.
39 I am borrowing 'investment in whiteness' from George Lipsitz, *The Possessive Investment in Whiteness* (Philadelphia: University of Temple Press, 2006).
40 Adams, *Education*, p. 1033.
41 Slotkin, *Gunfighter Nation*, p. 64; Adams, *Education*, p. 1035.
42 John Carlos Rowe, 'Adams's education in the age of imperialism', in Rowe (ed.), *New Essays on The Education of Henry Adams* (Cambridge: Cambridge University Press, 1996), p. 96.
43 Adams, *Education*, p. 1052.
44 Ibid., p. 1039.
45 Ibid., p. 1067.
46 Mark Seltzer, *Bodies and Machines* (New York and London: Routledge, 1992), p. 30.
47 Adams, *Education*, p. 1070.
48 Ibid.
49 Weinstein, 'From true woman', p. 309.
50 Adams, *Education*, p. 1075.
51 Jennifer L. Fleissner, *Women, Compulsion, Modernity: The Moment of American Naturalism* (Chicago: University of Chicago Press, 2004), p. 1.
52 Adams, *Education*, p. 1105.
53 Ibid., p. 1125.
54 Seltzer, *Bodies and Machines*, p. 30.
55 Adams, *Education*, pp. 1126–7.
56 Fleissner, *Women, Compulsion, Modernity*, p. 4.
57 Adams, *Education*, p. 1123.
58 Fleissner, *Women, Compulsion, Modernity*, p. 3.
59 Adams, *Education*, pp. 1131–2.
60 Martin, *Universe of Force*, p. 133.
61 Adams, *Education*, p. 1138.
62 Cary Wolfe, *Animal Rites: American Culture, the Discourse of Species, and Posthumanist Theory* (Chicago: University of Chicago Press, 2003), pp. 203–4.
63 Martin, *Universe of Force*, pp. 136, 137.
64 Ernest Samuels, *Henry Adams: The Major Phase* (Cambridge, MA: Harvard University Press, 1964), p. 393.
65 Adams, *Education*, p. 1154.

66 Ibid., pp. 1153, 1156.
67 Ibid., p. 1164.
68 Ibid., p. 1075.
69 Ibid., p. 1155.
70 Ibid., pp. 1163, 1164.
71 Ibid., p. 1167.
72 Rowe, 'Imperialism', p. 110.
73 Adams, *Education*, p. 1180.
74 Rowe, 'Imperialism', p. 107.
75 Ibid., p. 108.
76 Ibid., p. 109.
77 Ibid.
78 Ibid., p. 110.
79 Hayles, *Chaos Bound*, p. 88. While Hayles comments that '*The Education* marks a midpoint between Kelvin's mid-nineteenth-century vision of entropy as universal dissipation and [Claude] Shannon's reconceptualization of it as information in mid-twentieth century' (p. 61), she focuses mainly on examining Adams's images of chaos and order.
80 Adams, *Education*, p. 1156.
81 N. Katherine Hayles, *How We Became Posthuman: Virtual Bodies in Cybernetics, Literature, and Informatics* (Chicago: University of Chicago Press, 1999), p. 8.
82 Adams, *Education*, p. 1157.
83 Howard Horwitz, 'The *Education* and the salvation of history', in Rowe, *New Essays*, p. 119.
84 Adams, *Education*, p. 1158.
85 Ibid., pp. 1153, 1168.
86 Ibid., p. 1174.
87 Adams, *Education*, p. 1176; Hayles, *Posthuman*, p. 8.
88 Hayles, *Posthuman*, p. 8.
89 Horwitz, 'Salvation of history', p. 142.
90 Hayles, *Posthuman*, pp. 8, 67–8.
91 Ibid., p. 69.
92 In Wolfe, *Animal Rites*, pp. 204–5. For further discussion on the limits of Adams's vision, refer to Carolyn Porter, *Seeing and Being: The Plight of the Participant Observer in Emerson, James, Adams, and Faulkner* (Middletown, CT: Wesleyan University Press, 1981).
93 Adams, *Education*, p. 1133.
94 Rowe remarks: 'Adams advises the most powerful leaders of his age to pursue policies that would consolidate the twentieth-century economic

authority of the United States in international trade and serve as the [basis] for that special brand of US *neocolonialism* understood in the era of the Vietnam War in the phrase "establishing spheres of influence"'. See 'Imperialism', p. 89.

95 Pynchon, *Against the Day*, p. 10.

96 Fleissner, *Women, Compulsion, Modernity*, p. 1; Adams, *Education*, p. 1172.

97 Hayles, *Chaos Bound*, p. 87.

98 Horwitz, 'Salvation of history', p. 139.

99 Adams, *Education*, p. 1180.

100 Hayles, *Posthuman*, p. 13. Marvin Minsky, in particular, is critiqued for expressing the 'dream' of disembodiment through a suggestion that 'it will soon be possible to extract human memories from the brain and import them, intact and unchanged, to computer disks'.

101 Adams, *Education*, pp. 1173–5. Decker, in 'Disease of omniscience', finds Adams's remarks about a new evolutionary 'mind' credible, while Martin, in *Universe of Force*, sees them as further proof of Adams's irrationality.

102 Eric D. Schneider and Dorion Sagan, *Into the Cool: Energy Flow, Thermodynamics, and Life* (Chicago: University of Chicago Press, 2005), p. 238.

103 Adams, *Education*, pp. 1176, 1177; Levenson, 'Practicing history', p. 372.

104 Adams, *Education*, p. 1178.

105 Schneider and Sagan, *Into the Cool*, p. 6.

106 Horwitz, 'Salvation of history', p. 143.

107 Hayles, *Posthuman*, p. 102.

108 Ibid.

109 Ibid., p. 103.

110 Ibid., p. 109.

111 Ibid.

112 Erich Jantsch, *The Self-Organizing Universe: Scientific and Human Implications of the Emerging Paradigm of Evolution* (New York: Pergamon, 1980), p. 87.

113 Jeffrey S. Wicken, 'Thermodynamics, evolution, and emergence: ingredients for a new synthesis', in Bruce. H. Weber, David. J. Depew and James D. Smith (eds), *Entropy, Information, and Evolution* (Cambridge, MA and London: MIT Press, 1988), pp. 146, 145. For a critique of Wicken's argument, see Hayles, *Chaos Bound*, pp. 50–1.

114 Hayles, *Chaos Bound*, p. 80.

115 Niklas Luhmann, *Ecological Communication*, trans. John Bednarz, Jr (Chicago and London: University of Chicago Press, 1989), p. 16.

116 Schneider and Sagan, *Into the Cool*, p. 89.

117 Adams, *Education*, p. 1178.

118 Schneider and Sagan, *Into the Cool*, p. 91.

119 Ira Livingston, *Between Science and Literature: An Introduction to Autopoetics* (Urbana, IL and Chicago: University of Illinois Press, 2006), p. 138.

120 Henry Bumstead to Henry Adams, 16 June 1910, *Henry Adams Papers*, microfilm edition (Boston: Massachusetts Historical Society, 1984). Permission to quote granted by the MHS. Also see Martin, *Universe of Force*, pp. 142–3.

121 Martin, *Universe of Force*, p. 143.

122 Henry Adams, 'The Rule of Phase Applied to History', in *The Degradation of the Democratic Dogma* (New York: Peter Smith, 1949), p. 281.

123 William Jordy, *Henry Adams: Scientific Historian* (New Haven, CT: Yale University Press, 1952), p. 128.

124 Adams, 'Rule of Phase', p. 297.

125 Cary Wolfe, *What Is Posthumanism?* (Minneapolis: University of Minnesota Press, 2010), pp. 36–40.

126 Adams, 'Rule of Phase', p. 298.

127 Ibid., p. 299.

128 Ibid., p. 301, 304.

129 Adams, 'A Letter to American Teachers of History', in *Degradation*, p. 260.

130 Marginalia by Henry Adams in Hans Driesch, *The Science and Philosophy of the Organism, vol. II* (London: A. & C. Black, 1908), p. 339, Henry Adams Library, Massachusetts Historical Society.

131 Jane Bennett, *Vibrant Matter: A Political Ecology of Things* (Durham, NC and London: Duke University Press, 2010), pp. 73–4.

132 Bruce Clarke, 'Systems theory', in Bruce Clarke and Manuela Rossini (eds), *The Routledge Companion to Literature and Science* (New York and London: Routledge, 2011), p. 219.

133 A turn away from the future is, granted, reflected through the sheer repetition of Adams's argument (the two parts, 'The Problem' and 'The Solutions', being very similar), and the frequent, almost impulsive, recourse to a myriad of statistical evidence (that is rarely contextualized and, at times, is rather loosely related to his argument).

134 Adams, 'Letter', p. 158.

135 Adams, *Education*, p. 271.

136 Paul A. Bové, 'Henry Adams and the American System', in Decker and Harbert, *The Need to Know*, p. 1.

137 Ibid., p. 8.

138 Jordy, *Scientific Historian*, p. 207.

139 Adams, 'Letter', p. 184.

140 In Adams, 'Letter', pp. 188–9.

141 Adams, 'Letter', pp. 189–90.

142 Ibid., p. 208.

143 Vern Wagner, *The Suspension of Henry Adams: A Study of Manner and Matter* (Detroit: Wayne State University Press, 1969), p. 214.

144 Jordy, *Scientific Historian*, p. 210.

145 Wolfe, *Posthumanism*, p. 36.

146 Adams, 'Letter', p. 216.

147 Jacques Derrida, *Without Alibi*, ed. and trans. Peggy Kamuf (Stanford, CA: Stanford University Press, 2002), pp. 210, 72.

148 Ibid., p. 73.

149 In Wolfe, *Posthumanism*, p. 10.

150 Wolfe, *Posthumanism*, p. 6.

151 Ibid., p. 19.

152 Wolfe, *Animal Rites*, p. 200.

153 Wolfe, *Posthumanism*, p. 116.

154 Adams, 'Letter', p. 260. I am borrowing 'paranoid style' from the title of Richard Hofstadter's famous study, *The Paranoid Style in American Politics* (New York: Alfred A. Knopf, 1966). It seems likely that Adams's paranoia is refracted through his experience of the dynamos at the 1900 Paris Exposition. In a 7 November letter that same year to John Hay, he remarks that he liked to 'sit by the hour over the great dynamos, watching them run as noiselessly and as smoothly as the planets, and asking them – with infinite courtesy – where in Hell they are going . . . The Gods are not in it'. See J. C. Levenson et al. (eds), *The Letters of Henry Adams, vol. V: 1899–1905* (Cambridge, MA: Harvard University Press, 1988), p. 169. Note that Adams, moreover, appears to anticipate Derrida's concerns about language as a system, especially its 'machinelike automaticity', in Derrida, *Without Alibi*, p. 134. As Derrida then points out, 'the "I" itself' could be 'the "I" of anyone at all', what he then calls the 'usurpation of the singular *I* by the universal *I*, ineluctable substitution and subterfuge that makes all "justice" "unjust"'.

155 Adams, 'Letter', p. 263.

156 William Rasch, 'Immanent systems, transcendental temptations, and the limits of ethics', in William Rasch and Cary Wolfe (eds), *Observing Complexity: Systems Theory and Postmodernity* (Minneapolis: University of Minnesota Press, 2000), p. 91.

157 Ibid., p. 93.

158 Ibid.
159 Wolfe, *Animal Rites*, p. 200.
160 Adams, 'Letter', p. 245.
161 Bové, 'American System', p. 14.

Conclusion

1 Paul A. Bové, 'Henry Adams and the American System', in William M. Decker and Earl N. Harbert (eds), *Henry Adams and the Need to Know* (Boston: Massachusetts Historical Society, 2005), p. 2.
2 John Carlos Rowe, 'Adams's education in the age of imperialism', in Rowe (ed.), *New Essays on the Education of Henry Adams* (Cambridge: Cambridge University Press), p. 101.
3 Ibid., p. 89.
4 Bové, 'American System', p. 14.
5 Bové, 'American System', p. 14; Henry Adams, *The Education of Henry Adams*, in *Adams: Novels, Mont Saint Michel, The Education*, ed. Ernest Samuels and Jayne N. Samuels (New York: Library of America, 1983), p. 1101. Later in this chapter, Adams's own performance will be addressed.
6 Bové, 'American System', p. 14.
7 Ibid., pp. 2–3.
8 Adams, *Education*, p. 1174.
9 Bové, 'American System', p. 8.
10 Henry Adams, 'A Letter to American Teachers of History', in *The Degradation of the Democratic Dogma* (New York: Peter Smith, 1949), p. 158. The idea of 'flooding the market' was discussed at the business meeting of the Environment and Culture Caucus, Conference of the American Studies Association, San Juan, Puerto Rico, November 2012.
11 Ibid., p. 216.
12 For a recent example, see Barbara Demick, 'China hit by extreme air pollution', *Los Angeles Times* (13 January 2013), *www.latimes.com/nationworld/world/la-fg-china-air-pollution-20130114,0,1184.story*, accessed 18 January 2013.
13 Stacy Alaimo, *Bodily Natures: Science, Environment, and the Material Self* (Bloomington and Indianapolis: Indiana University Press, 2010), p. 16.
14 Adams, *Education*, p. 1174.
15 Ursula K. Heise, *Sense of Place and Sense of Planet: The Environmental Imagination of the Global* (Oxford: Oxford University Press, 2008), p. 146.

16 William Rasch, *Niklas Luhmann's Modernity: The Paradoxes of Differentiation* (Stanford, CA: Stanford Press, 2006), p. 24.
17 Heise, *Sense of Place*, p. 147.
18 In Heise, *Sense of Place*, p. 151.
19 Heise, *Sense of Place*, p. 152; Rasch, *Luhmann's Modernity*, p. 33.
20 Rasch, *Luhmann's Modernity*, p. 82.
21 Niklas Luhmann, *Ecological Communication*, trans. John Bednarz, Jr (Chicago: University of Chicago Press, 1989), p. 14.
22 Rasch, *Luhmann's Modernity*, p. 102.
23 Andrew McMurry, *Environmental Renaissance: Emerson, Thoreau, and the Systems of Nature* (Athens: University of Georgia Press, 2003), p. 213.
24 Luhmann, *Ecological Communication*, p. 19.
25 McMurry, *Environmental Renaissance*, p. 228.
26 Heise, *Sense of Place*, p. 151.
27 Bové, 'American System', p. 15.
28 Thomas Pynchon, *Against the Day* (New York: Penguin, 2006), p. 10.
29 J. C. Levenson et al. (eds), *The Letters of Henry Adams, vol. V: 1899–1905* (Cambridge, MA: Harvard University Press, 1988), p. 678.
30 Bové, 'American System', p. 13.
31 Heise, *Sense of Place*, p. 147.
32 Ibid., p. 149.
33 Rasch, *Luhmann's Modernity*, p. 122.
34 Heise, *Sense of Place*, p. 149.
35 Adams, *Education*, pp. 937, 1163–4.
36 Adams, 'Letter', p. 260.
37 Bové, 'American System', p. 7.
38 Adams, *Education*, p. 1070.
39 Cindy Weinstein, 'From true woman to new woman to virgin', in Decker and Harbert, *The Need to Know*, p. 309.
40 Rowe, 'Imperialism', p. 109.
41 Jennifer L. Fleissner, *Women, Compulsion, Modernity: The Moment of American Naturalism* (Chicago: University of Chicago Press, 2004), p. 190.
42 Ira Livingston, *Between Science and Literature: An Introduction to Autopoetics* (Urbana and Chicago: University of Illinois Press, 2006), p. 140.
43 Lee Smolin, in Livingston, *Science and Literature*, p. 137.
44 Theodore Dreiser, *Sister Carrie*, in *Sister Carrie, Jennie Gerhardt, Twelve Men*, ed. Richard Lehan (New York: Library of America, 1987), p. 447.
45 Dreiser, *Sister Carrie*, p. 435.
46 Bové, 'American System', p. 5.

47 Thanks to Nihad Farooq for suggesting this idea.

48 As Donna J. Haraway proposes, '[C]yborg politics insist on noise and advocate pollution, rejoicing in the illegitimate fusions of animal and machine. These are the couplings which make Man and Woman so problematic, subverting the structure of desire, the force imagined to generate language and gender, and so subverting the structure and modes of reproduction of "Western" identity, of nature and culture, mirror and eye, slave and master, body and mind'. Refer to *Simians, Cyborgs, and Women: The Reinvention of Nature* (New York and London: Routledge, 1991), p. 176.

49 Jacques Derrida, *Without Alibi*, ed. and trans. Peggy Kamuf (Stanford, CA: Stanford University Press, 2002), p. 210.

50 Bové, 'American System', 14.

51 Judith Butler, *Bodies That Matter: On the Discursive Limits of 'Sex'* (New York and London: Routledge, 1993), pp. 12, 226.

52 Dreiser, *Sister Carrie*, p. 438.

53 Cary Wolfe, *What Is Posthumanism?* (Minneapolis: University of Minnesota Press, 2010), p. 237.

54 Recall the scene earlier discussed in chapter 2. Drouet and Hurstwood's desire for Carrie, while she is acting a role during *Under the Gaslight*, is amplified and expanded: 'All the gentlemen yearned toward her'. See Dreiser, *Sister Carrie*, p. 403.

55 Livingston, *Science and Literature*, p. 138.

56 Dreiser, *Sister Carrie*, 442.

57 Nancy Bentley, *The Ethnography of Manners: Hawthorne, James, Wharton* (Oxford: Oxford University Press, 2003), p. 174.

58 Edith Wharton, *The Custom of the Country*, in *Edith Wharton: Novels*, ed. R. W. B. Lewis (New York: Library of America, 1985), p. 966.

59 Bentley, *Ethnography of Manners*, p. 183. See also my discussion in chapter 4, section III.

60 Ibid., pp. 179, 203.

61 Wharton, *Custom of the Country*, p. 779.

62 Cary Wolfe, *Animal Rites: American Culture, the Discourse of Species, and Posthumanist Theory* (Chicago: University of Chicago Press, 2003), p. 202.

63 Wharton, *Custom of the Country*, p. 688.

64 Jacques Derrida, *Specters of Marx*, trans. Peggy Kamuf (New York and London: Routledge, 1994), p. 169. For a discussion of 'spectral life', see Eugene Thacker, *After Life* (Chicago and London: University of Chicago Press, 2010), pp. 249–57.

65 Wharton, *Custom of the Country*, pp. 802–3.

66 Bentley, *Ethnography of Manners*, pp. 167–72, 194.
67 Livingston, *Science and Literature*, p. 142.
68 James Heintz and Radhika Balakrishnan, 'Debt, power, and crisis: social stratification and the inequitable governance of financial markets', *American Quarterly*, Special Issue: Race, Empire, and the Crisis of the Subprime, 64/1 (September 2012), 399, 400.
69 Wharton, *Custom of the Country*, p. 1014.
70 Levenson et al., *Letters of Henry Adams*, p. 678.
71 Bové, 'American System', p. 15.
72 Weinstein is an important exception. Refer to 'From true woman', p. 309.
73 Bové, 'American System', p. 14.
74 Adams, 'Letter', p. 260.
75 Alaimo, *Bodily Natures*, p. 21.
76 Karen Barad, *Meeting the Universe Halfway: Quantum Physics and the Entanglement of Matter and Meaning* (Durham, NC and New York: Duke University Press, 2007), pp. 145–6.
77 Michael J. Watts, 'The tale of two gulfs: life, death, and dispossession along two oil frontiers', *American Quarterly*, Special Issue: Race, Empire, and the Crisis of the Subprime, 64/1 (September 2012), 440.
78 Ibid., 448.
79 Cary Wolfe, *Before the Law: Humans and Other Animals in a Biopolitical Frame* (Chicago and London: University of Chicago Press, 2013), p. 90.
80 Ibid., p. 91.
81 In Bové, 'American System', p. 12. Bové elaborates: 'Cooper's writings reflect quite an old but mainstream position within what Edward W. Said has called Orientalism, although Cooper has nicely updated the policy rhetoric'. Furthermore, Bové asserts that Adams would be against such a policy.
82 Watts, 'Two gulfs', 464.
83 Alaimo's recent work looks at the destructive effects of oceanic pollution on global ecosystems. See 'States of suspension: trans-corporeality at sea', *Interdisciplinary Studies in Literature and the Environment* (*ISLE*), 19/3 (Summer 2012), 476–93. Also refer to Wolfe, *Before the Law*, p. 105.
84 Jane Bennett, *Vibrant Matter: A Political Ecology of Things* (Durham, NC and London: Duke University Press, 2010), p. 49.
85 Ursula K. Heise, 'Who is responsible for the environment?' (paper presented at the Conference of the Modern Language Association, Boston, MA, January 2013). Permission to cite paper granted by author.
86 Wolfe, *Before the Law*, pp. 59–60.
87 Bennett, *Vibrant Matter*, p. 122.

88 Barad, *Universe*, p. 391.

89 Alaimo, *Bodily Natures*, p. 146.

90 Wolfe, *Posthumanism*, pp. 63, 81.

91 Ibid., p. 81.

92 Ibid., p. 88.

93 Ibid., p. 39.

94 Jacques Derrida, 'The animal that therefore I am (more to follow)', trans. David Willis, *Critical Inquiry*, 28/2 (Winter 2002), 401.

95 Jacques Derrida, 'And say the animal responded?', trans. David Willis, in Cary Wolfe (ed.), *Zoontologies: The Question of the Animal* (Minneapolis: University of Minnesota Press, 2003), p. 137.

96 Jacques Derrida, '"Eating well", or the calculation of the subject: an interview with Jacques Derrida', in Eduardo Cadava, Peter Connor and Jean-Luc Nancy (eds), *Who Comes after the Subject?* (New York and London: Routledge, 1991), p. 116.

97 Wolfe, *Posthumanism*, p. 89.

98 Ibid., p. 90.

99 Ibid., p. 118.

100 Levenson et al. (eds), *The Letters of Henry Adams*, p. 701.

101 Pynchon, *Against the Day*, p. 10.

102 Wolfe, *Before the Law*, pp. 45–6.

103 Ibid., p. 45. As Wolfe remarks, Derrida 'pulls no punches in criticizing "this violence that some would compare to the worst cases of genocide", a genocide made even more perverse by the fact that millions of animals are "exterminated *by means of their continued existence or even their overpopulation*"' (p. 44). At the same time, Wolfe points out, 'Derrida (an Algerian Jew) is well aware of the complexities of the analogy here, of course, and he reminds us that "one should neither abuse the figure of genocide nor too quickly consider it explained away"'. Thus Derrida, Wolfe states, helps us to see 'some important' ways that the Nazi concentration camps differ from factory farming (*Before the Law*, pp. 44–5).

104 Ibid., p. 37. It should be noted, as Wolfe calls to our attention, that Giorgio Agamben's stance here is somewhat extreme, namely for how his argument tends to downplay the critical issue of historical context in Foucault's biopolitical thinking.

105 Wolfe, *Before the Law*, p. 43.

106 Derrida, 'The animal that therefore I am', 398.

107 Wolfe, *Before the Law*, p. 9.

108 Wolfe, *Before the Law*, p. 55; Giorgio Agamben, *Homo Sacer: Sovereign Power and Bare Life*, trans. Daniel Heller-Roazen (Stanford, CA: Stanford

University Press, 1998); Derrida, 'The animal that therefore I am', 415. Wolfe is particularly critical of Roberto Esposito's 'insistence on "the principle of unlimited equivalence for every single form of life"' (*Before the Law*, p. 56). As we have already seen, Esposito's view appears to differ little from – and to have the same problems as – Deep Ecology thinking of the 1970s and 1980s.

109 Jacques Derrida, *The Beast and the Sovereign, vol. I*, ed. Michel Lisse, Marie-Louise Mallet and Ginette Michaud, trans. Geoffrey Bennington (Chicago: University of Chicago Press, 2009), p. 183.

110 Wolfe, *Before the Law*, p. 67.

111 Derrida, 'The animal that therefore I am', 398.

112 Wolfe, *Before the Law*, p. 78.

113 Ibid.

114 Ibid., pp. 82, 83. Elsewhere Wolfe provocatively states that 'we' are, always/already, subjected 'to the alterity and radical otherness of time'. Refer to *Posthumanism*, pp. 298–9.

115 Ibid., p. 84.

116 Ibid., p. 90.

117 Ibid., p. 91.

118 Ibid., p. 92.

119 Ibid., p. 95.

120 Ibid. Wolfe earlier remarks: '[I]t is all the more instructive to recall, as Derrida points out in his essay, "Before the Law," that when Freud addresses the problem of the origin of law (what is its basis? on what moral foundations does it rest?), he resorts to what amounts to a sacrifice of the animal and, more broadly, of animality, as the means by which both the human and the basis of the law are secured' (p. 9). Wolfe explains elsewhere that, for Derrida, 'the specific practice of eating animals becomes simply one more version of the larger symbolic structure by which "Man" in the Western philosophical tradition secures its transcendence through mastery of nature, repression of the body – everything that Derrida associates with the term "carnophallogocentrism"'. Wolfe goes on to discuss the political limitations of Derrida's views. See *Posthumanism*, pp. 95–8.

121 In Wolfe, p. 98.

122 Wolfe, *Before the Law*, p. 98. What forms a crucial dimension of Wolfe's argument here is his referencing of Esposito's discussion of Adolf Hitler's 'telegram number 71 sent from his bunker in Berlin as the Allies were approaching'; the telegram 'ordered that all means of subsistence for the German people, who had shown themselves unworthy of the Reich, should be destroyed'. Framing Wolfe's argument is the debate

about synthetic meat that demonstrates the prevalence of the motif of sacrifice in thinking about sustainability that would, for Wolfe, explain why 'many of the most prominent advocates of sustainable agriculture actually *oppose* synthetic meat' (p. 97).

123 Ibid., p. 99.

124 Ibid., p. 101.

125 Ibid. China's situation indicates that 'the very ecological sustainability of the planet is at stake in the repression of this violence against non-human animals'. Wolfe then quotes from Paul Roberts to propose that the decision of whether to eat meat has moved 'from the category of lifestyle choice to that of collective responsibility'.

126 Ibid., pp. 8–9.

127 Ibid., p. 9.

128 Wolfe, *Posthumanism*, p. 82; *Before the Law*, p. 24. Also see Michael Lundblad, *The Birth of a Jungle: Animality in Progressive-Era US Literature and Culture* (Oxford: Oxford University Press, 2013).

129 Earle Labor, Robert C. Leitz III and I. Milo Shepard (eds), *The Letters of Jack London, vol. I: 1896–1905* (Stanford, CA: Stanford University Press, 1988), p. 89.

130 London, *White Fang*, in *Jack London: Novels and Stories*, ed. Donald Pizer (New York: Library of America, 1982), p. 234.

131 Vicki Hearne, *Adam's Task: Calling Animals by Name* (New York: Random House, 1987), p. 48.

132 Wolfe, *Before the Law*, p. 67.

133 London, *White Fang*, p. 269.

134 Ibid., p. 182.

135 Ibid., p. 204.

136 Tellingly, the narrator states that Gray Beaver did not 'sound' the 'deeps in [White Fang's] nature', which distinguishes Gray Beaver from Weedon Scott. See *White Fang*, p. 187. The implication here is that the wolf can learn love *only* from a white, upper-class American.

137 Charles Darwin, *The Expression of Emotions in Man and Animals*, ed. Francis Darwin (London: William Pickering, 1989), p. 215.

138 Mark Seltzer, *Bodies and Machines* (New York and London: Routledge, 1992), p. 169.

139 London, *White Fang*, pp. 275–6.

140 Ibid., p. 280.

141 Ibid., p. 268.

142 Ibid., p. 283.

143 Ibid., p. 92.

144 Seltzer, *Bodies and Machines*, p. 166.
145 Thacker, *After Life*, p. 234.
146 Wolfe, *Before the Law*, pp. 80–1.
147 Wolfe, *Posthumanism*, p. 85.
148 London, 'The Other Animals', in Dale L. Walker and Jeanne Campbell Reesman (eds), *No Mentor But Myself: Jack London on Writing and Writers*, 2nd edn (Stanford, CA: Stanford University Press, 1999), p. 120.
149 Wolfe, *Before the Law*, p. 103. For an example of 'autoimmune consequences', refer to Wolfe's discussion of Hitler's telegram number 71 (p. 98).
150 Ibid.
151 Ibid.

Epilogue

1 Henry Adams, 'A Letter to American Teachers of History' in *The Degradation of the Democratic Dogma* (New York: Peter Smith, 1949), p. 260.
2 Michel Foucault, *'Society Must Be Defended': Lectures at the Collège de France, 1975–1976*, ed. Mauro Bertani and Alessandro Fontana, trans. David Macey (New York: Picador, 2003), p. 255.
3 Gavin Jones, *American Hungers: The Problem of Poverty in US Literature, 1840–1845* (Princeton, NJ and Oxford: Princeton University Press, 2008), p. 85.
4 Edith Wharton, *The House of Mirth*, in *Edith Wharton: Novels*, ed. R. W. B. Lewis (New York: Library of America, 1985), p. 347. Jennifer L. Fleissner moves us towards the language of systems (and systems theory) when she remarks that 'the word' affirms 'such a moment's necessary impossibility', that is, the 'impossibility' of Lily and Selden 'connected at last'. Fleissner also points out, quoting from Wai Chee Dimock, that the destruction of the letters signifies an 'erasure from history'. See Jennifer L. Fleissner, *Women, Compulsion, Modernity: The Moment of American Naturalism* (Chicago: University of Chicago Press, 2004), p. 195.
5 Cary Wolfe, *Before the Law: Humans and Other Animals in a Biopolitical Frame* (Chicago and London: University of Chicago Press, 2013), pp. 100, 103.
6 London, *White Fang*, in *Jack London: Novels and Stories*, ed. Donald Pizer (New York: Library of America, 1982), pp. 277, 279.
7 Wolfe, *Before the Law*, p. 103. The idea, as Wolfe puts it, that a legal decision 'is always already traced with the automaticity and mechanicity of a reaction' makes it possible to examine this idea through Derrida's

discussion of the interrelationship between promise and threat. Refer to Jacques Derrida, *Without Alibi*, ed. and trans. Peggy Kamuf (Stanford, CA: Stanford University Press, 2002), p. 155.

8 Wolfe, *Before the Law*, p. 105. Also see p. 11 for how Wolfe's depiction of 'a newly expanded community of the living' is framed by 'two examples of how non-human animals are currently framed at opposite extremes in relation to moral standing and legal protection'. His first example is: 'On June 25 of 2008, the Environmental Committee of the Spanish Parliament approved resolutions to grant basic rights to Great Apes on a quite traditional model of human rights'. His second example is: 'According to statistics provided by the US Department of Agriculture, in the previous year, 2007, about nine billion animals were killed in the United States for food – the vast majority of them raised in Confined Animal Feeding Operations (CAFOs) or "factory farms" – double the number in 1980'.

BIBLIOGRAPHY

Adams, Carol J., 'The feminist traffic in animals', in Greta Gaard (ed.), *Ecofeminism: Women, Animals, Nature* (Philadelphia: Temple University Press, 1993), pp. 197–206.

Adams, Henry, 'A Letter to American Teachers of History', in Adams, *The Degradation of the Democratic Dogma* (New York: Peter Smith, 1949), pp. 137–263.

Adams, Henry, 'The Rule of Phase Applied to History', in Adams, *The Degradation of the Democratic Dogma* (New York: Peter Smith, 1949), pp. 267–311.

Adams, Henry, *The Education of Henry Adams*, in *Adams: Novels, Mont Saint Michel, The Education*, ed. Ernest Samuels and Jayne N. Samuels (New York: Library of America, 1983), pp. 717–1181.

Adams, John Quincy, 'Science versus art-appreciation', *Popular Science Monthly*, 62/25 (March 1903), 453–62.

Agamben, Giorgio, *Homo Sacer: Sovereign Power and Bare Life*, trans. Daniel Heller-Roazen (Stanford, CA: Stanford University Press, 1998).

Alaimo, Stacy, *Undomesticated Ground: Recasting Nature as Feminist Space* (Ithaca, NY and London: Cornell University Press, 2000).

Alaimo, Stacy, *Bodily Natures: Science, Environment, and the Material Self* (Bloomington and Indianapolis: Indiana University Press, 2010).

'American trade with Central and South American countries', *Scientific American*, 78/19 (7 May 1898), 291.

Ammons, Elizabeth, 'Edith Wharton and the issue of race', in Millicent Bell (ed.), *The Cambridge Companion to Edith Wharton* (Cambridge: Cambridge University Press, 1995), pp. 68–86.

Angell, Norman, *The Great Illusion: A Study of the Relation of Military Power to National Advantage*, 4th edn (New York and London: G. P. Putnam's Sons, 1913).

Auerbach, Jonathan, *Male Call: Becoming Jack London* (Durham, NC: Duke University Press, 1996).

Bannister, Robert C., *Social Darwinism: Science and Myth in Anglo-American Social Thought* (Philadelphia: Temple University Press, 1979).

Barad, Karen, *Meeting the Universe Halfway: Quantum Physics and the Entanglement of Matter and Meaning* (Durham, NC and London: Duke University Press, 2007).

Bauer, Dale M., *Edith Wharton's Brave New Politics* (Madison: University of Wisconsin Press, 1994).

Bauer, Dale M., '"In the blood": sentiment, sex, and the ugly girl', *differences*, 11/3 (1999), 57–75.

Bauer, Dale M., 'Wharton's "others": addiction and intimacy', in Carol. J. Singley (ed.), *A Historical Guide to Edith Wharton* (Oxford and New York: Oxford University Press, 2003), pp. 115–45.

Beck, Ulrich, *Risk Society: Towards a New Modernity*, trans. Mark Ritter (London: Sage, 1992).

Bederman, Gail, *Manliness and Civilization: A Cultural History of Gender and Race in the United States, 1880–1917* (Chicago: University of Chicago Press, 1995).

Bender, Bert, *The Descent of Love: Darwin and the Theory of Sexual Selection in American Fiction, 1871–1926* (Philadelphia: University of Pennsylvania Press, 1996).

Beniger, James, *The Control Revolution: Technological and Economic Origins of the Information Society* (Cambridge, MA: Harvard University Press, 1986).

Bennett, Jane, *Vibrant Matter: A Political Ecology of Things* (Durham, NC and London: Duke University Press, 2010).

Bensaude-Vincent, Bernadette, 'A historical perspective on science and its "Others"', *Isis*, 100/2 (June 2009), 259–68.

Benstock, Shari, *No Gifts from Chance: A Biography of Edith Wharton* (New York: Scribner, 1994).

Bentley, Nancy, *The Ethnography of Manners: Hawthorne, James, Wharton* (Cambridge: Cambridge University Press, 1995).

Bentley, Nancy, 'Wharton, travel, and modernity', in Carol. J. Singley (ed.), *A Historical Guide to Edith Wharton* (Oxford and New York: Oxford University Press, 2003), pp. 147–79.

Berkove, Lawrence I., 'Jack London and evolution: from Spencer to Huxley', *American Literary Realism*, 36/3 (Spring 2004), 243–55.

Berlant, Lauren, *The Anatomy of National Fantasy: Hawthorne, Utopia, and Everyday Life* (Chicago: University of Chicago Press, 1991).

Bové, Paul A., 'Henry Adams and the American System', in William Merrill Decker and Earl N. Harbert (eds), *Henry Adams and the Need to Know* (Boston: Massachusetts Historical Society, 2005), pp. 1–17.

Bowler, Peter J., *The Non-Darwinian Revolution* (Baltimore, MD: Johns Hopkins University Press, 1988).

Brinton, Daniel G., 'The aims of anthropology', *Popular Science Monthly*, 47/1 (November 1895), 59–72.

Bruni, John, 'Becoming American: evolution and performance in Edith Wharton's *The Custom of the Country*', *Intertexts*, 9/1 (Spring 2005), 43–59.

Bruni, John, 'Furry logic: biological kinship and empire in Jack London's *The Call of the Wild*', *Interdisciplinary Studies in Literature and Environment* (*ISLE*), 14/1 (Winter 2007), 25–49.

Bruni, John, 'Performing the perfect dog: the reconstruction of gender in Jack London's *The Call of the Wild* and *White Fang*', in Bernadette H. Hyner and Precious McKenzie Stearns (eds), *Forces of Nature: Natural(izing) Gender and Gender(ing) Nature in the Discourses of Western Culture* (Newcastle upon Tyne: Cambridge Scholars Press, 2009), pp. 174–210.

Bruni, John, 'The miseducation of Henry Adams: fantasies of race, citizenship, and Darwinian dynamos', in Jeanette Eileen Jones and Patrick B. Sharp (eds), *Darwin in Atlantic Cultures: Evolutionary Visions of Race, Gender, and Sexuality* (New York and London: Routledge, 2010), pp. 260–82.

Butler, Judith, *Bodies That Matter: On the Discursive Limits of 'Sex'* (New York and London: Routledge, 1993).

Butler, Judith, *The Psychic Life of Power: Theories in Subjection* (Stanford, CA: University of Stanford Press, 1997).

Cattell, James McKeen, 'The progress of science: the popularization of science', *Popular Science Monthly*, 72/19 (April 1908), 382–4.

Clarke, Bruce and Mark B. N. Hansen, 'Introduction: neocybernetic emergence', in Clarke and Hansen (eds), *Emergence and Embodiment: New Essays on Second-Order Systems Theory* (Durham, NC: Duke University Press, 2009), pp. 1–25.

Clarke, Bruce, 'Systems theory', in Bruce Clarke and Manuela Rossini (eds), *The Routledge Companion to Literature and Science* (New York and London: Routledge, 2011), pp. 214–25.

Conder, John J., *Naturalism in American Fiction: The Classic Phase* (Lexington: University of Kentucky Press, 1984).

Coolidge, Archibald Cary, *The United States as a World Power* (New York: Macmillan, 1909).

Coulter, John M., 'Public interest in research', *Popular Science Monthly*, 72/19 (August 1905), 306–12.

Darwin, Charles, *On the Origin of Species: A Facsimile of the First Edition* (Cambridge, MA: Harvard University Press, 1966).

Darwin, Charles, *The Descent of Man and Selection in Relation to Sex*, 2 vols (New York: D. Appleton & Company, 1871).

Darwin, Charles, *The Descent of Man and Selection in Relation to Sex* (New York: AMS Press, 1972).

Darwin, Charles, *The Descent of Man and Selection in Relation to Sex*, 2nd edn (Chicago: Rand, McNally & Co., 1974).

Darwin, Charles, *The Expression of Emotions in Man and Animals*, ed. Francis Darwin (London: William Pickering, 1989).

Davis, Mike, *Ecology of Fear: Los Angeles and the Imagination of Disaster* (New York: Metropolitan Books, 1998).

Decker, William M., 'A martyr to the disease of omniscience', in William Merrill Decker and Earl N. Harbert (eds), *Henry Adams and the Need to Know* (Boston: Massachusetts Historical Society, 2005), pp. 318–44.

Deloria, Philip J., *Playing Indian* (New Haven, CT and London: Yale University Press, 1998).

Demick, Barbara, 'China hit by extreme air pollution', *Los Angeles Times* (13 January 2013), *latimes.com/nationworld/world/la-fg-china-air-pollution-20130114,0,1184.story*, accessed 18 January 2013.

Depew, David J. and Bruce H. Weber (eds), *Darwinism Evolving: Systems Dynamics and the Genealogy of Natural Selection* (Cambridge, MA and London: MIT Press, 1995).

Derrida, Jacques, '"Eating well", or the calculation of the subject: an interview with Jacques Derrida', in Eduardo Cadava, Peter Connor, and Jean-Luc Nancy (eds), *Who Comes after the Subject?* (New York and London: Routledge, 1991), pp. 96–119.

Derrida, Jacques, *Specters of Marx*, trans. Peggy Kamuf (New York and London: Routledge, 1994).

Derrida, Jacques, 'The animal that therefore I am (more to follow)', trans. David Willis, *Critical Inquiry*, 28/2 (Winter 2002), 369–418.

Derrida, Jacques, *Without Alibi*, ed. and trans. Peggy Kamuf (Stanford, CA: Stanford University Press, 2002).

Derrida, Jacques, 'And say the animal responded?', trans. David Wills, in Cary Wolfe (ed.), *Zoontologies: The Question of the Animal* (Minneapolis: University of Minnesota Press, 2003), pp. 121–46.

Derrida, Jacques, *The Beast and the Sovereign, vol. I*, ed. Michel Lisse, Marie-Louise Mallet and Ginette Michaud, trans. Geoffrey Bennington (Chicago: University of Chicago Press, 2009).

Diamond, Elin, 'Introduction', in Elin Diamond (ed.), *Performance and Cultural Politics* (New York and London: Routledge, 1996), pp. 1–13.

Dimock, Wai Chee, 'Debasing exchange: Edith Wharton's *The House of Mirth*', *PMLA*, 10/5 (1985), 783–92.

Dreiser, Theodore, *Sister Carrie*, in *Sister Carrie, Jennie Gerhardt, Twelve Men*, ed. Richard Lehan (New York: Library of America, 1987), pp. 3–455.

Driesch, Hans, *The Science and Philosophy of the Organism, vol. II* (London: A. & C. Black, 1908).

Dwight, Eleanor, 'Wharton and art', in Carol. J. Singley (ed.), *A Historical Guide to Edith Wharton* (Oxford and New York: Oxford University Press, 2003), pp. 181–210.

Dyke, C., 'Cities as dissipative structures', in Bruce H. Weber, David J. Depew and James D. Smith (eds), *Entropy, Information, and Evolution: New Perspectives on Physical and Biological Evolution* (Cambridge, MA: MIT Press, 1988), pp. 355–67.

Etulain, Richard, *Beyond the Missouri: The Story of the American West* (Albuquerque: University of New Mexico Press, 2006).

Fleissner, Jennifer L., *Women, Compulsion, Modernity: The Moment of American Naturalism* (Chicago: University of Chicago Press, 2004).

Foucault, Michel, 'Right of death and power over life', in Paul Rabinow (ed.), *The Foucault Reader* (New York: Pantheon, 1984), pp. 259–72.

Foucault, Michel, *'Society Must Be Defended': Lectures at the Collège de France, 1975–1976*, ed. Mauro Bertani and Alessandro Fontana, trans. David Macey (New York: Picador, 2003).

Franklin, W. S., 'The Second Law of Thermodynamics: its basis in intuition and common sense', *Popular Science Monthly*, 76/15 (March 1910), 227–45.

Fullerton, William Morton, *Problems of Power: A Study of International Politics from Sadowa to Kirk-Kilisse* (London: Constable & Company Ltd, 1913).

Gaard, Greta, 'Living interconnections with animals and nature', in Gaard (ed.), *Ecofeminism: Women, Animals, Nature* (Philadelphia: Temple University Press, 1993), pp. 1–12.

Gair, Christopher, *Complicity and Resistance in Jack London's Novels* (Lewiston, NY: Edwin Mellon, 1997).

Gair, Christopher, 'The crumbling structure of "appearances": representation and authenticity in the *House of Mirth* and the *Custom of the Country*', *Modern Fiction Studies*, 43/2 (1997), 349–73.

Garber, Marjorie, *Dog Love* (New York: Simon & Schuster, 1996).

Gilman, Charlotte Perkins, *Woman and Economics* (New York: Harper & Row, 1966).

Goldner, Ellen J., 'The lying women and the cause of social anxiety: interdependence and the woman's body in *The House of Mirth*', *Women's Studies*, 21 (1992), 285–305.

Gould, Stephen Jay, *Ever Since Darwin* (New York: W. W. Norton, 1977).

Gould, Stephen Jay, *The Panda's Thumb: More Reflections in Natural History* (New York and London: W. W. Norton, 1980).

Hamilton, David Mike, *'The Tools of My Trade': The Annotated Books in Jack London's Library* (Seattle: University of Washington Press, 1986).

Haraway, Donna J., *Simians, Cyborgs, and Women: The Reinvention of Nature* (London and New York: Routledge, 1991).

Haraway, Donna J., *Modest_Witness@Second Millenium. FemaleMan_Meets_OncoMouse* (New York: Routledge, 1997).

Haraway, Donna J., *The Companion Species Manifesto: Dogs, People, and Significant Otherness* (Chicago: Prickly Paradigm Press, 2003).

Harley, Louis R., 'Race mixture and national character', *Popular Science Monthly*, 47/1 (May 1895), 86–91.

Harrington, M. W., 'Natural products and resources of the Philippine Islands', *Scientific American*, 78/23 (4 June 1898), 355.

Hartman, Saidiya V., *Scenes of Subjection: Terror, Slavery, and Self-Making in Nineteenth-Century America* (New York and Oxford: Oxford University Press, 1997).

Hayles, N. Katherine, *Chaos Bound* (Ithaca, NY and London: Cornell University Press, 1990).

Hayles, N. Katherine, *How We Became Posthuman: Virtual Bodies in Cybernetics, Literature, and Informatics* (Chicago: University of Chicago Press, 1999).

Hearne, Vicki, *Adam's Task: Calling Animals by Name* (New York: Random House, 1987).

Hedrick, Joan, *Solitary Comrade: Jack London and His Work* (Chapel Hill: University of North Carolina Press, 1982).

Heintz, James and Radhika Balakrishnan, 'Debt, power, and crisis: social stratification and the inequitable governance of financial markets', *American Quarterly*, Special Issue: Race, Empire, and the Crisis of the Subprime, 64/1 (September 2012), 387–409.

Heise, Ursula K., *Sense of Place and Sense of Planet: The Environmental Imagination of the Global* (Oxford: Oxford University Press, 2008).

Heise, Ursula K., 'Who is responsible for the environment?' (paper presented at Conference of the Modern Language Association, Boston, MA, January 2013).

Hofstadter, Richard, *Social Darwinism in American Thought* (Boston: Beacon Press, 1955).

Hofstadter, Richard, *The Paranoid Style in American Politics and Other Essays* (New York: Alfred A. Knopf, 1966).

Hopkins, Lisa, 'Jack London's evolutionary hierarchies: dogs, wolves, and men', in Lois A. Cuddy and Claire M. Roche (eds), *Evolution and Eugenics in American Literature and Culture, 1880–1940: Essays on Ideological Conflict and Complicity* (Lewisburg, PA: Bucknell University Press, 2003), pp. 89–101.

Horwitz, Howard, 'The *Education* and the salvation of history', in John Carlos Rowe (ed.), *New Essays on The Education of Henry Adams* (Cambridge: Cambridge University Press, 1996), pp. 115–56.

Howard, June, *Form and History in American Literary Naturalism* (Chapel Hill: University of North Carolina Press, 1985).

Howard, Maureen, '*The House of Mirth*: the bachelor and the baby', in Millicent Bell (ed.), *The Cambridge Companion to Edith Wharton* (Cambridge and New York: Cambridge University Press, 1995), pp. 137–56.

Howerth, Ira W., 'The classification of the arts', *Popular Science Monthly*, 70/25 (May 1907), 429–36.

Huxley, Thomas Henry, *Evolution and Ethics: The Romanes Lecture* (New York: D. Appleton & Company, 1894).

Jantsch, Erich, *The Self-Organizing Universe: Scientific and Human Implications of the Emerging Paradigm of Evolution* (New York: Pergamon, 1980).

Jesse, Edward, *Anecdotes of Dogs* (London: Bell & Daldy, 1873).

Johnson, Michael L., *Hunger for the Wild: America's Obsession with the Untamed West* (Lawrence: University of Kansas Press, 2007).

Jones, Gavin, *American Hungers: The Problem of Poverty in US Literature, 1840–1945* (Princeton, NJ and Oxford: Princeton University Press, 2008).

Kaplan, Amy, 'Nation, region, and empire', in Emory Elliot (ed.), *Columbia History of the American Novel* (New York: Columbia University Press, 1991), pp. 240–67.

Kassanoff, Jennie A., *Edith Wharton and the Politics of Race* (Cambridge: Cambridge University Press, 2004).

Kasson, John F., *Houdini, Tarzan, and the Perfect Man: The White Male Body and The Challenge of Modernity in America* (New York: Hill and Wang, 2001).

Keefer, Karen M., 'Dirty water and clean toilets: medical aspects of the 1904 Louisiana Purchase Exposition', *Gateway Heritage*, 9/1 (Summer 1988), 32–7.

Keller, Evelyn Fox, 'From secrets of life to secrets of death', in Keller et al. (eds), *Three Cultures: Fifteen Lectures on the Confrontation of Academic Cultures* (The Hague: Universitaire Pers Rotterdam, 1989), pp. 3–17.

Kellogg, Vernon L., *Darwinism Today: A Discussion of Present-Day Scientific Criticism of the Darwinian Selection Theories* (New York: Henry Holt and Co., 1908).

Kolodny, Annette, *The Lay of the Land: Metaphor as Experience and History in American Life and Letters* (Chapel Hill: University of North Carolina Press, 1975).

Labor, Earle, Robert C. Leitz III and I. Milo Shepard (eds), *The Letters of Jack London, vol. I: 1896–1905* (Stanford, CA: Stanford University Press, 1988).

Lacan, Jacques, 'The subversion of the subject and the dialectic of desire in the Freudian unconscious', trans. Alan Sheridan, in *Écrits: A Selection* (New York and London: Norton, 1977), pp. 292–325.

LaFollette, Marcel C., *Making Science Our Own: Public Images of Science, 1910–1955* (Chicago and London: University of Chicago Press, 1990).

Lagemann, Robert T., 'New light on old rays: N-rays', *American Journal of Physics*, 45/3 (March 1977), 281–4.

Levenson, J. C., 'Henry Adams, U. S. Grant, and evolution: practicing history in the age of Darwin', in William Merrill Decker and Earl N. Harbert (eds), *Henry Adams and the Need to Know* (Boston: Massachusetts Historical Society, 2005), pp. 345–74.

Lewis, R. W. B., *Edith Wharton* (New York: Harper & Row, 1975).

Lipsitz, George, *The Possessive Investment in Whiteness* (Philadelphia: University of Temple Press, 2006).

Livingston, Ira, *Between Science and Literature: An Introduction to Autopoetics* (Urbana and Chicago: University of Illinois Press, 2006).

Livingston, Ira, 'Complex visuality: the radical middleground', in Bruce Clarke and Mark B. N. Hansen (eds), *Emergence and Embodiment: New Essays on Second-Order Systems Theory* (Durham, NC: Duke University Press, 2009), pp. 246–62.

Lock, Robert H., *The Recent Progress in the Study of Variation, Heredity, and the Environment* (London: John Murray, 1906).

London, Jack, 'All Gold Canyon', in *Jack London: Novels and Stories*, ed. Donald Pizer (New York: Library of America, 1982), pp. 777–96.

London, Jack, *The Call of the Wild*, in *Jack London: Novels and Stories*, ed. Donald Pizer (New York: Library of America, 1982), pp. 5–86.

London, Jack, 'The Other Animals', in Dale L. Walker and Jeanne Campbell Reesman (eds), *No Mentor But Myself: Jack London on Writing and Writers*, 2nd edn (Stanford, CA: Stanford University Press, 1999), pp. 108–20.

London, Jack, *White Fang*, in *Jack London: Novels and Stories*, ed. Donald Pizer (New York: Library of America, 1982), pp. 91–284.

Luhmann, Niklas, *Ecological Communication*, trans. John Bednarz, Jr (Chicago: University of Chicago Press, 1989).

Luhmann, Niklas, *Art as a Social System*, trans. Eva M. Knodt (Stanford, CA: Stanford University Press, 2000).

Lundblad, Michael, *The Birth of a Jungle: Animality in Progressive-Era US Literature and Culture* (Oxford: Oxford University Press, 2013).

Mallea, Amahia, 'Downstreamers: public health and relationships on the Missouri River', *Agricultural History*, 76/2 (Spring 2002), 393–404.

Margulis, Lynn and Dorion Sagan, *What is Life?* (New York: Simon & Schuster, 1995).

Margulis, Lynn and Dorion Sagan, *Acquiring Genomes: A Theory of the Origin of Species* (New York: Basic Books, 2002).

Martin, Jay, *Harvests of Change: American Literature, 1865–1914* (Englewood Cliffs, NJ: Prentice Hall, 1967).

Martin, Ronald E., *American Literature and the Universe of Force* (Durham, NC: Duke University Press, 1981).

Maturana, Humberto R. and Francisco J. Varela, *The Tree of Knowledge: The Biological Roots of Human Understanding*, rev. edn, trans. Robert Paolucci, foreword by J. Z. Young (Boston: Shambhala, 1992).

Mayr, Ernst, 'Introduction', in Darwin, *On the Origin of Species: A Facsimile of the First Edition* (Cambridge, MA: Harvard University Press, 1966), pp. vii–xxvii.

McMurry, Andrew, *Environmental Renaissance: Emerson, Thoreau, and the Systems of Nature* (Athens: University of Georgia Press, 2003).

Merleau-Ponty, Maurice, *Phenomenology of Perception*, trans. Colin Smith (London: Routledge & Kegan Paul, 1962).

Michaels, Walter Benn, *The Gold Standard and the Logic of Naturalism: American Literature at the Turn of the Century* (Berkeley: University of California Press, 1987).

Michaels, Walter Benn, *Our America: Nativism, Modernism, and Pluralism* (Durham, NC: Duke University Press, 1997).

Michaud, Gustave, 'The climate of our new possessions', *Scientific American*, 83/11 (15 September 1900), 171.

Mitchell, Lee Clark, *Determined Fictions: American Literary Naturalism* (New York: Columbia University Press, 1989).

Moore, John Howard, *Universal Kinship* (Chicago: Charles H. Kerr & Co., 1906).

Morton, Timothy, *Ecology Without Nature: Rethinking Environmental Aesthetics* (Cambridge, MA: Harvard University Press, 2009).

Munn, Orson Desaix, 'Political crisis', *Scientific American*, 73/26 (28 December 1895), 402–3.

'The N-rays: are they real or illusory?', *Scientific American*, 90/23 (4 June 1904), 434.

Nourie, Alan and Barbara Nourie (eds), *N. W. Ayer & Son's American Newspaper Annual and Directory* (Philadelphia: N. W. Ayer & Son, 1895–1905).

Nourie, Alan and Barbara Nourie (eds), *American Mass-Market Magazines* (New York and Westport, CT: Greenwood Press, 1990).

Orlando, Emily J., 'Picturing Lily: body art in The House of Mirth', in Gary Totten (ed.), *Memorial Boxes and Guarded Interiors: Edith Wharton and Material Culture* (Tuscaloosa: University of Alabama Press, 2007), pp. 83–110.

Osborn, Henry Fairfield, 'Life and works of Darwin', *Popular Science Monthly*, 74 (April 1909), 315–43.

'Our trade with Africa', *Scientific American*, 74/25 (20 June 1896), 391.

Patterson, Martha H., 'Incorporating the new woman in Wharton's *The Custom of the Country*', *Studies in American Fiction*, 26/2 (1998), 213–36.

Pearson, Karl, *The Grammar of Science*, 2nd edn (London: Adam and Charles Black, 1900).

Pettegrew, John, *Brutes in Suits: Male Sensibility in America, 1890–1920* (Baltimore, MD: Johns Hopkins University Press, 2007).

'The Philippine Islands', *Scientific American*, 78/19 (7 May 1898), 290–1.

Pizer, Donald, *The Theory and Practice of American Literary Naturalism: Selected Essays and Reviews* (Carbondale: Southern Illinois University Press, 1982).

Pizer, Donald (ed.), *The Cambridge Companion to American Realism and Naturalism: Howells to London* (Cambridge: Cambridge University Press, 1995).

Porter, Carolyn, *Seeing and Being: The Plight of the Participant Observer in Emerson, James, Adams, and Faulkner* (Middletown, CT: Wesleyan University Press, 1981).

Preston, Claire, *Edith Wharton's Social Register* (New York: St. Martin's Press, 2000).

Pynchon, Thomas, *Slow Learner* (Boston and Toronto: Little, Brown & Company, 1984).

Pynchon, Thomas, *Against the Day* (New York: Penguin, 2006).

Rasch, William, 'Immanent systems, transcendental temptations, and the limits of ethics', in William Rasch and Cary Wolfe (eds), *Observing Complexity: Systems Theory and Postmodernity* (Minneapolis: University of Minnesota Press, 2000), pp. 73–98.

Rasch, William, *Niklas Luhmann's Modernity: The Paradoxes of Differentiation* (Stanford, CA: Stanford University Press, 2006).

Robertson, Roland, *Globalization: Social Theory and Global Culture* (London: Sage, 1992).

Roosevelt, Theodore, *American Ideals and Other Essays, Social and Political, vol. I* (New York and London: G. P. Putnam's Sons, 1907).

Roosevelt, Theodore, 'The Strenuous Life', in *The Works of Theodore Roosevelt, vol. XV* (New York: Scribner's, 1923–6), pp. 270–1.

Rosaldo, Renato, *Culture and Truth: The Remaking of Social Analysis* (Boston: Beacon Press, 1989).

Rowe, John Carlos, 'Adams's education in the age of imperialism', in Rowe (ed.), *New Essays on The Education of Henry Adams* (Cambridge: Cambridge University Press, 1996), pp. 87–114.

Samuel, T. D., Jr, 'The water supply system of Kansas City, Missouri', *American Water Works Association Journal*, 22/9 (1930), 1236–46.

Samuels, Ernest, *Henry Adams: The Major Phase* (Cambridge, MA: Harvard University Press, 1964).

Sapora, Carol, 'Undine Spragg, the mirror and the lamp in *The Custom of the Country*', in Gary Totten (ed.), *Memorial Boxes and Guarded Interiors: Edith Wharton and Material Culture* (Tuscaloosa: University of Alabama Press, 2007), pp. 265–86.

Sassoubre, Ticien Marie, 'Property and identity in *The Custom of the Country*', *Modern Fiction Studies*, 49/4 (2003), 687–713.

Schneider, Eric D. and Dorion Sagan, *Into the Cool: Energy Flow, Thermodynamics, and Life* (Chicago: University of Chicago Press, 2005).

Schweighauser, Philipp, *The Noises of American Literature, 1890–1985: Toward a History of Literary Acoustics* (Gainesville: University of Florida Press, 2006).

Secord, James A., 'Knowledge in transit', *Isis*, 95/4 (December 2004), 654–72.

Sedgwick, Eve Kosofsky, *Tendencies* (Durham, NC: Duke University Press, 1993).

Seltzer, Mark, *Bodies and Machines* (New York and London: Routledge, 1992).

Shank, Barry, 'Iterating the collective subject: performative history and the ethics of cultural politics' (unpublished paper delivered at the Performance Seminar, Hall Center for the Humanities, University of Kansas, Autumn 1998).

Shapiro-Shapin, Carolyn G., '"A really excellent scientific contribution": scientific creativity, scientific professionalism, and the Chicago Drainage Case, 1900–1906', *Bulletin of the History of Medicine*, 71/3 (1997), 385–411.

Sherow, James E., 'The art of water and the art of living', *Kansas History*, 25/1 (2002), 52–71.

Slotkin, Richard, *Gunfighter Nation: The Myth of the Frontier in Twentieth-Century America* (New York: Athenaeum, 1992).

Smith, Crosbie and Ian Higginson, '"No traces of a beginning, no prospect of an end": Henry Adams, Charles Lyell, and the politics of uniformity', in William Merrill Decker and Earl N. Harbert (eds), *Henry Adams and the Need to Know* (Boston: Massachusetts Historical Society, 2005), pp. 104–137.

Spencer, Herbert, 'Professional institutions. IV. Orator and poet, actor and dramatist', *Popular Science Monthly*, 47/35 (August 1895), 433–45.

Spencer, Herbert, 'Professional institutions. XI. Painter', *Popular Science Monthly*, 48/41 (March 1896), 697–704.

Stanley, Henry M., 'The development of Africa', *Scientific American*, 75/16 (17 October 1896), 298.

Streeby, Shelley, *American Sensations: Class, Empire, and the Production of Popular Culture* (Berkeley, Los Angeles and London: University of California Press, 2002).

Sundquist, Eric J. (ed.), *American Realism: New Essays* (Baltimore, MD: Johns Hopkins University Press, 1982).

Tabbi, Joseph, 'World systems colliding: Thomas Pynchon and Niklas Luhmann', in Hannes Bergthaller and Carsten Schinko (eds), *Addressing Modernity: Social Systems Theory and US Cultures* (Amsterdam and New York: Rodopi, 2011), pp. 313–42.

Thacker, Eugene, *After Life* (Chicago and London: University of Chicago Press, 2010).

Thacker, Eugene, 'Necrologies, or death of the "body politic"', in Patricia T. Clough and Craig Willse (eds), *Beyond Biopolitics* (Durham, NC and New York: Duke University Press, 2011), pp. 139–62.

Thorndike, Edward L., 'The evolution of the human intellect', *Popular Science Monthly*, 50/60 (November 1901), 58–65.

Tichi, Cecelia, 'Emerson, Darwin, and *The Custom of the Country*', in Carol. J. Singley (ed.), *A Historical Guide to Edith Wharton* (Oxford and New York: Oxford University Press, 2003), pp. 89–114.

Tobey, Ronald C., *The American Ideology of National Science, 1919–1930* (Pittsburgh, PA: University of Pittsburgh Press, 1971).

Topham, Jonathan, 'Focus: historicizing "popular science"', *Isis*, 100/2 (June 2009), 310–18.

Trachtenberg, Alan, *Shades of Hiawatha: Staging Indians, Making Americans, 1880–1930* (New York: Hill and Wang, 2004).

Wagner, Vern, *The Suspension of Henry Adams: A Study of Manner and Matter* (Detroit: Wayne State University Press, 1969).

Walcutt, Charles C., *American Literary Naturalism: A Divided Stream* (Minneapolis: University of Minnesota Press, 1956).

Wald, Priscilla, *Constituting Americans: Cultural Anxiety and Narrative Form* (Durham, NC: Duke University Press, 1995).

Wallace, Alfred Russel, *Darwinism: An Exposition of the Theory of Natural Selection with Some of Its Applications* (London: Macmillan, 1901).

Wallerstein, Immanuel, *World Systems Analysis: An Introduction* (Durham, NC and London: Duke University Press, 2004).

Watts, Michael J., 'The tale of two gulfs: life, death, and dispossession along two oil frontiers', *American Quarterly*, Special Issue: Race, Empire, and the Crisis of the Subprime, 64/1 (September 2012), 437–67.

Webster, Gordon, 'America's intellectual product', *Popular Science Monthly*, 72/13 (March 1908), 193–210.

Wegener, Frederick, '"Rabid imperialist": Edith Wharton and the obligations of empire in modern American fiction', *American Literature*, 72/4 (December 2000), 783–812.

Weinstein, Cindy, 'From true woman to new woman to virgin', in William Merrill Decker and Earl N. Harbert (eds), *Henry Adams and the Need to Know* (Boston: Massachusetts Historical Society, 2005), pp. 300–14.

Wharton, Edith, *A Backwards Glance* (New York: D. Appleton-Century, 1934).

Wharton, Edith, *The Custom of the Country*, in *Edith Wharton: Novels*, ed. R. W. B. Lewis (New York: Library of America, 1985), pp. 623–1014.

Wharton, Edith, *The House of Mirth*, in *Edith Wharton: Novels*, ed. R. W. B. Lewis (New York: Library of America, 1985), pp. 3–347.

Whipple, George C., 'Municipal water-works laboratories', *Popular Science Monthly* (December 1900), 172–82.

Wicken, Jeffrey S., 'Thermodynamics, evolution, and emergence: ingredients for a new synthesis', in Bruce H. Weber, David J. Depew and James D. Smith (eds), *Entropy, Information, and Evolution: New Perspectives on Physical and Biological Evolution* (Cambridge, MA: MIT Press, 1988), pp. 139–69.

Williams, Raymond, *Marxism and Literature* (Oxford and New York: Oxford University Press, 1977).

Wolfe, Cary, *Critical Environments: Postmodern Theory and the Pragmatics of the 'Outside'* (Minneapolis and London: University of Minnesota Press, 1998).

Wolfe, Cary, *Animal Rites: American Culture, the Discourse of Species, and Posthumanist Theory* (Chicago and London: University of Chicago Press, 2003).

Wolfe, Cary, *What Is Posthumanism?* (Minneapolis and London: University of Minnesota Press, 2009).

Wolfe, Cary, *Before the Law: Humans and Other Animals in a Biopolitical Frame* (Chicago and London: University of Chicago Press, 2013).

Young, Egerton R., *My Dogs in the Northland*, 5th edn (New York: Fleming H. Revell & Co., 1902).

Zanine, Louis J., *Mechanism and Mysticism: The Influence of Science on the Thought and Work of Theodore Dreiser* (Philadelphia: University of Pennsylvania Press, 1993).

INDEX